Black & White and Read All Over

A Collection of BarbWire Columns

By Barbara R. Lumley

First Edition, Second Printing, 2020.

ISBN 9781734525908

Copyright © 2019, Steven W Lumley
All Rights Reserved. No part of this book may be reproduced or transmitted in any form by any means, electronic or mechanical, including photocopying, recording or by any other information storage or retrieval system, without the prior written permission of the publisher, except for the inclusion of brief quotations in a review.
All photos are the property of the author and publisher and are protected by copyright.

Published by Steven W. Lumley
3059 Mahogany Run Cir NW
North Canton, OH 44720, USA
(330) 433-0301

DISCLAIMER

This material represents the opinions of the author and should be treated as such. Despite the best efforts of all concerned, typographical or editorial errors may occur. The author and publisher make no warranty, express or implied, nor assume any legal liability for any data that has been mistakenly included or omitted, added or deleted, or for the accuracy, completeness, or usefulness of any other information contained herein. The publisher and author disclaim any liability, in whole or in part, arising from information contained in the book. Any resemblance to actual events, locales, or persons, living or dead, short or tall, religious or atheist, is entirely coincidental. Read at your own risk. Do not attempt to read this book while driving or operating a cellular device. No people or animals were harmed in the making of this book. The mention of commercial products, their source, or their use in connection with material reported herein is not to be construed as an actual or implied endorsement of such products. Always seek the advice of your physician or other qualified health provider for medical conditions. The author and publisher advise readers to take full responsibility for their safety and know their limits.

Foreword

Barb was born into the rural farm life in 1939. She was the daughter of George and Laura Wagner, though she represented the fifth generation on the family farm that had supported the family since 1840. She was not one to play around the house, as Laura always remarked about her "messing up the rugs", so she paddled after her father, George, and became his right-hand man. Barb grew up tending to all the chores required of farm life at that time.

In the early 1950s, George bought a Registered Holstein cow at a local auction to try to improve the quality of the herd. Registered Holsteins would transform their herd and Barb's future. Over the next fifty-six years, Barb developed a small, quality registered herd of Holsteins with quality genetics. This was her focus and passion. She was involved in the Ohio Holstein community at a time when few women were in leadership or decision-making positions on a farm. She helped build proven progeny with show-quality animals that would be well-represented in the showring at both the state and national level. Offspring of the foundation of her herd would be merchandised in public and private sales. She had spent the entirety of her life working on the farm daily, solving problems, building a first-class herd of Registered Holsteins, and contributing to the local, state, and national organizations for the Holstein breed.

Back in 2002, Barb was asked to write a regular column for the Ohio Holstein News. Thus began her writing career. When the herd was sold in 2009, she could focus her attention on a life-long love of hers, a deep desire to be a writer.

My grandmother, Laura, always took the opportunity to bestow upon us grandkids the value of good penmanship. During our elementary years, if we were sitting on the porch with her or sitting under the maple tree enjoying a cool summer night breeze, my Grandmother would talk about how she always practiced writing each of the cursive letters and the admiration of beautiful penmanship. It is clear that she imprinted this value on my mother as well, for my mother has beautiful flowing cursive penmanship, complete with gracefully rounded letters and long tails on each appropriate letter and word.

Her writing process also reflects the love of the physical part of writing. She drafts each piece on a yellow legal pad, fully writing and editing by hand before it ever sees the light of the computer monitor. She may have multiple columns in process at any given time. Some come together very quickly, others take days or weeks to reach a workable state.

"I have an idea for a column!" is a usual discussion topic when I go to visit. Sometimes I will send her interesting articles that will lead to an idea for a column. She has enjoyed conjuring up ideas for a column, making observations in life, and weaving them into a publishable piece, and polishing the piece with just the right words. Most of all, she enjoys sharing her writing with others, and especially hearing feedback on a particular memory, or how a piece touched someone.

This collection represents a portion of her writing the "BarbWire" column over the past eleven years. I have pulled together some of the best pieces that reflect on life and I hope you enjoy sharing these columns from my mother.

— Steven Wagner Lumley, son

BARBWIRE

Is there life after milking dairy cows? I'm sitting here at the computer at 5:30 A.M. seeking the answer to that question. My cows and bred heifers left for their new home at Velvet View Farms on May 10. However, I am "easing my way out" instead of going "cold turkey". I am still milking two cows. Glitter remained because of her 14 years of age, and her Red and White daughter, Ringafire-Red, is recovering from a foot injury. So I have discovered that milking two cows is not that different from milking twenty five. You still have to go to the barn twice a day, get the cows in, feed, milk, clean the stable, and wash everything up. It just doesn't take as long!

Since I don't have enough milk to ship, we have to do something with it, and dumping good milk down the drain is hard to do. So, Don is eating a lot of potato soup, graham cracker pudding, ice cream, etc., and we are giving milk to neighbors, relatives, and strangers on the street. I am making lots of ice cream, my sister-in-law is churning butter, and we deliver a couple times a week to a refrigerator in a friend's garage, where his friends gather and help themselves to milk, soda, or "other beverages". So far, the milk is disappearing just as fast as the "other beverages".

Everyone keeps asking what I am doing with all my spare time. There are plenty of things that I need to catch up on, especially house cleaning. I have always managed to keep things fairly clean and neat, but I am not the housekeeper that my Mom was. She always kept things perfect when I was growing up, crocheted doilies covered every table and the backs of the couch and chairs, curtains were starched and perfectly hung. Everything had its place and heaven help you if you messed things up! So I learned at an early age to go help my Dad. The dogs and calves became my friends, playmates, and siblings. Thus my interest in cows was developed. Being on the lead strap with a beautiful cow attached brings me a lot more joy than a dust rag in my hand!

I am enjoying announcing at District shows and County fairs and visiting with all the people there. It's always a pleasure to watch those beautiful heifers and cows parade the show ring. And it is a joy to see so many young people working with dairy cattle and developing a passion for them. I find myself spending more time on the computer and learning how to do more with it, even though it gives me a headache at times! And, of course, being involved with the Ohio Holstein Association keeps me busy and is always a pleasure.

I often wonder what my life would have been like if there hadn't been a rule in Carroll County that said, "Girls must own a registered calf in order to join a 4-H livestock club".

I should thank the "male chauvinist" who made that rule, as he changed my life forever. My first Registered Holstein was purchased and my passion for showing and breeding them began. If

FIRST REGISTERED 4H CALF

you are going to go to all the work of milking dairy cows two or three times a day, seven days a week, three hundred sixty five days a year, you might as well reap the rewards that come with owning and breeding registered dairy cattle, for there are many. If you like milking cows, go for it! Do I miss my cows? Of course I do, just like every other registered breeder who has had to make this decision. I miss being able to look out my

bathroom window early in the morning and see the cows coming to the barn from the pasture. Such a pretty picture at sunrise on a June morning. Little things still get to me, like seeing the grass growing up in the path the cows walked to the barn, or "Moose" walking into the show ring at the Red and White Show at State Fair. Do I still get tears in my eyes when I think about my cows? You betcha. Would I do it all over again? In a heartbeat!

BARBWIRE

My first recollection of our dairy herd was of an assortment of Guernsey, Brown Swiss, Milking Shorthorn, and various crosses of those breeds. My parents were just glad to have a few cows of any caliber to milk and help provide some income. Every farmer kept a bull running with the herd. Ours was a Guernsey, weighing about 900 pounds, and one that I will never forget, as he almost killed my Dad. Thank God my Uncle Bob just happened to be here that evening when the cows came in to be milked. I was too little and could never have saved him. I remember screaming and running to bury my face in the side of one of the cows and crying as they fought the bull. The sight of my Dad afterwards will forever be emblazoned in my memory. Blood running down his face, clothes torn off, but fortunately no bones were broken. Two days later I would hear my Dad say that there wasn't a spot on his body bigger than a dime that wasn't black and blue, and it would be several days before he could walk without pain, especially downhill!

The bull was sent to the local auction as soon as possible, leaving us with no way to breed our cows. We had just recently bought our first Holstein bull, but he was just a baby. In those days if you didn't have your own bull, you relied on your closest neighbor who did. Thus later on Brownie and Herfie would enter the herd! Brownie was Brown Swiss, Guernsey, Milking Shorthorn crossed with the neighbor's Hereford bull. Herfie was Guernsey-Hereford cross. Brownie milked pretty well, but Herfie didn't! It was surprising though how good their udders were! Their genomics would certainly have been an interesting study.

Holsteins were becoming very popular and very high in

price. We would beef a couple cows and attend a farm sale hoping to buy one, only to come home empty handed, as we didn't have enough money. Finally Dad went to a sale and bought our first "registered Holstein". Before he loaded her, the farmer picked up a nice round stone and placed it under the tailgate. After she was loaded he handed the stone to my Dad, telling him "Place the stone under the tailgate before you unload her and she will never get homesick". My Dad did just as he was instructed and she settled in perfectly. Of course, we had to endure the joke from the "colored breeders" about how you could drop a quarter into the bottom of a bucket of Holstein milk and still read "In God we trust" on it! We named her "Pride" and I can still see her, mostly black, medium size, wide flat rump and a nicely attached udder with teats just the right size for hand milking. She would be with us for several

years, milking well and giving us several heifer calves. There was just one problem. We never received the registration papers in spite of several letters to the previous owner.

It would be 1953 before we would finally own a real registered Holstein, when we purchased my first two calves for 4-H. They were sired by Pabst Burke Dell and I can still remember how thrilled I was when I finally received those registration papers in the mail. A year later we would return to the same Holstein breeder to purchase Melody Carnation Inka Fobes, sired by Carnation Cedric Inspiration, who would go on to be the foundation for most of our registered Holsteins. She would milk well, test 4%, classify Very Good 87, and give us several heifer calves. However, I still remember the disappointment when she calved with a beautiful heifer calf by Shanghigh Exec Double Bud that had black encircling her front leg and running down to touch the hoof. She couldn't be registered! The rules were no black touching the hoof, no black on the belly, no black hairs in the switch, and no red color. I told my Dad to send her to the community sale. I couldn't stand to own such a beautiful heifer calf with no registration papers. Thank goodness the Holstein Association finally abolished those rules, but not before good genetics were lost.

As an owner of a registered Holstein, I became a junior member of the National Holstein Association and later on a lifetime member. We became members of the Ohio Holstein Association. That would lead me to the show ring and I loved showing Holsteins! I would work my "fanny" off just to be allowed to go to the county fair or a Black and White Show. I was a member of the Stark-Carroll Holstein Club and attended my first State Show at

Lawrence McCullough's farm, now Paradise Valley Farms. Showing led us to Jim Lewis, who would encourage us to take a good calf to the State Show and State Fair, resulting in All-Ohio Awards. When districts were formed, I became active in District 3. Then Cal Wilcox came calling, looking for sale consignments for the Ohio Holstein Sales and insisting on my best. It was hard to let the good ones go, but being a small herd, we needed extra income. Those consignments meant roofing the barn, new equipment, and buying semen from top bulls. We attended the state conventions and became more involved in the Ohio Holstein Association.

Over the years we have traveled to many places and met many people involved with registered Holsteins. We have been given so many memories and so many friendships that we treasure. As I sit here today, writing a column for the Ohio Holstein News, I cannot help but wonder, "What if I had never owned anything but a grade cow?" That first registration paper and my first registered Holstein was life changing for me and my family. What could registration papers do for you?

WHAT MAKES A HOME HAPPY?

The answer to that question may surprise you. For some, the answer may be a big house, lots of money, fancy cars, exotic vacations, diamonds, designer clothes, etc. For many, those things play no part in making their home happy. Happiness can be many things.

It can be a special picture of an old barn, that hangs on the wall, painted by your friend, Marguerite. She was such fun, a great cook, loved Jersey cows, and her beagle dog, Andy. It might be the afghan you pull over yourself when you feel chilly, crocheted by your Mom's best friend and neighbor for sixty years. It could be the little nightstand sitting by your bed that belonged to your Grandpa. That was where he always kept his chewing tobacco. It was the one thing you asked for when he passed away.

Perhaps it is the smell of sheets dried on the clothesline on a breezy, spring day or the basket of clean white towels, neatly folded and ready to put away. It could be putting that extra warm quilt on the bed, as the snowflakes drift slowly down outside the window. Maybe it's sitting on the porch in the rocking chair, watching a soft, warm rain fall on the corn and hay fields, or the bunch of purple petunias blooming beside the front door. It could be that first red ripe tomato from the garden you have been weeding and tending. Or the pantry shelves and freezer filled with produce from that garden to enjoy through the winter.

What about the card and letter from the old friend you went to school with, who is now living far away. It could be sitting and reading a good book or listening to your favorite music. Everyone is happy when your child gets an A on a test, when they usually have to settle for B's or C's. Could it be the pictures of your grandchildren hanging on the wall, or the angel plaque they gave you for Christmas?

How about that new recipe you tried that didn't turn out the way it was supposed to, but your husband ate it anyway and told you it was good. It could be your Mom's Christmas cactus that has bloomed with beautiful red flowers every year since she passed away, in spite of the fact that you usually managed to kill every other plant you ever tried to grow!

Happiness could be the Excellent Holstein cows you have bred and the Grand Champion Banner that one of them won at the district show, and the trophies your children won at the State Fair. It might be your Black Lab laying his head on your knee and wagging his tail, or the calico cat curled up in your lap as you watch TV. Did you see that gorgeous pheasant rooster strutting his stuff out in the yard? Have you been watching all those hummingbirds zipping around the feeder on the front porch? Look around you. There is happiness in every home. You just have to know where to look for it!

WHAT ARE YOU FEEDING YOUR BRAIN?

If you are reading my column, the words are helping to keep your brain healthy. According to a recent article, The Perfect Brain Food, in Reader's Digest, "Reading isn't just filling your head --- it's nourishing it. The cheapest, easiest, and most time-tested way to sharpen your brain is right in front of you". It is called reading. As those of you who read my columns know, I have often mentioned my love for reading and encourage others to do the same.

Reading encourages the brain to work harder and better. For example if you read the following sentence, "The hiker stopped suddenly when the huge bull appeared on the path". Your brain must go to work to create the picture of the words you are reading. Reading gives your brain a unique pause for comprehension and insight. While you are not the one facing the bull, your brain acts as if you are and several parts of your brain go to work. Is the hiker calm or scared? Is it a man or woman? Is the hiker wearing red? (Bulls are not supposed to like red!) Is this a big bull or a little one? What color is this bull? What breed is that bull? (Is that really important!) Is the path wide or narrow? Are there any trees nearby that you can climb? Until you read on, your brain must work to create those images. If you are watching that scene on television or at the movies your brain does not have to work, thus it doesn't exercise. Without the sustained exercise of our reading "muscles" the brain loses its ability to control the intricate processes that allow us to read deeply.

I know several people who do very little reading. Some are retired and don't go out much during this time of year. They complain of the long, dark, dreary, dull days and some are depressed. Reading can help to pass the time, it can take you places you have never been before and it can be educational and very enjoyable. There is so much reading material of all kinds available and many libraries to loan books and reading material to you. Many of those same people worry about Alzheimer's disease. More and more information is out there suggesting that reading, doing crosswords puzzles, and other activities that make your brain work are very beneficial.

I have recently discovered that I am having trouble remembering phone numbers! Why? It's not just due to aging. Thanks to modern technology all I have to do to call someone is bring up the name and number on my phone and press a button! What if I am out somewhere and lose my phone or if something causes my phone to be destroyed? If I can't remember the number how do I contact someone? Are our young children, who all have phones these days, memorizing important phone numbers in case of emergency? 911 isn't always available. Are we all becoming too dependent on the modern technology for everything?

Hours can fly by when I get involved in Holstein magazines, especially the old ones. I enjoy novels, especially ones set in Old England (a favorite writer was the late Catharine Cookson) or during the Civil War. I subscribe to several magazines other than the farm ones I get, and several newspapers. So I seldom run out of reading material. I can always find something in each of them that is enjoyable or educational. There is reading material available on every subject you can think of and more.

As I write my columns, I hope that I am writing something that the readers can enjoy, something that they can learn from or just something that can help to make their day a little brighter. It is also satisfying to learn that my words are helping to keep my readers healthy. On these dark, dreary and snowy days, just like the words a bath product once advertised (remember Calgon?), "Reading --- take me away!"

STATE OF EMERGENCY – MY COMPUTER IS NOT WORKING

I need a fix! Whoa, wait a minute! Don't go down that wrong road, what I need is a computer fix ---my computer is NOT WORKING! Hard to imagine---just a few short years ago I didn't have a computer and didn't want one. Circumstances dictated that I had to get one and learn how to use it. Now here I am totally overwhelmed with frustration and anxiety because my computer isn't working! How did I get to this place in life?

Because I am a little "older" and wasn't raised with computers and all this technical stuff, I have trouble understanding how all this works. When I have a problem I have to call my eldest son. He is very knowledgeable in many areas including computers. He uses all that "computer lingo", does his best to explain, and I still don't get it. Thank goodness he is very patient with me! Unfortunately, today he is probably all tied up in problems with the company he runs, so he hasn't called me back.

The last few days have been very busy, as there is a lot going on with the Ohio Holstein Association. The Convention and Annual Meeting will be held March 10-11, 2017. We are working on the Convention Sale, March 10 at 11:00 at the Holmes County Fairgrounds. The catalog will soon be available and features 75 outstanding consignments, several with +2700 and up genomics. There is advertising to be done and details to be completed. Emails have been flying back and forth (or whatever they do) among all of us involved. The Ohio Holstein News will be coming out soon and I have been working with the new editor, Melissa Hart. Melissa is the editor of

Dairy Agenda Today on the internet, writes her column for several magazines and newspapers, and is a speaker. Every morning I click onto Dairy Agenda Today to read the latest news in the dairy industry and now I can't do that. It is a part of my morning routine --- right after I start the coffee perking! I have two columns that I need to send out and I can't do that! I know there are several email messages there, but I can't get to them to open them and see what they are about!

    I live down here among the hills where our internet service is dependent on towers ---no cable service here—and we need more towers! I have no cell phone service here unless I hop on my lawn mower and drive up on top of one of my hills. Text messages usually come in when I am driving into town and they can be days late. We recently had many gas and oil workers in my area from out of state. When they had a problem and needed to call someone, their cell phones wouldn't work, so they had to knock on someone's door and ask to use their landline phone. It could be a little unnerving when a couple big guys were sitting at my kitchen table speaking in Spanish and the only words I could remember from high school Spanish class were "como esta usted", "buenos dias" and "taco"!

    After a phone call, which involved a recording with a list of numbers to push (after listening to 7 or 8 of them I couldn't remember which one I needed to push so had to listen all over again) talking to four live people, each sending me on to someone else, finally the last one was a helpful one who explained the problem. Evidently I have used up all my "megabytes" (whatever those are). I have a limit and when it is used up I can't get on the internet. No one bothered to let me know I was running out! If I want to increase my limit it will cost more money (why doesn't that surprise me) or I can just wait until the limit is renewed later in the month. NOT AN OPTION! I have work to do.

When my phone starts ringing in the morning with people wanting to know what is wrong and why I am not emailing, at least I will be talking to people who have a good reason for calling, instead of telemarketers, gambling surveys, and recordings by someone wanting to take care of the warranty on my car, help with my hearing and funeral expenses, or help me if I fall and can't get up! It is hard to believe I actually miss my computer! It just seems life was so much simpler and easier when messages were written in pen and ink and went by Pony Express. Of course, even in those days the success of the ride depended on a good horse!

TECHNOLOGY IS GREAT WHEN IT WORKS

All this "new fangled" technology is great but when it doesn't work it can ruin your entire day! The Ohio Holstein Convention Sale is fast approaching, March 11 at Dover, and emails have been flying back and forth among those of us involved. Decisions on catalogs, added information, and ads for the sale needed to be made. Copies were being sent as attachments to emails and all was going well until suddenly none of the attachments on my email would open. I was being asked to proof read and send additional information, however I couldn't open anything so I could look at it. Along with all the sale information were new pictures of my great-granddaughter who is soon to celebrate her first birthday and I couldn't open those either!

I sent an email back to one of the newspapers we are advertising in expressing my dilemma. In return they asked if they could send a fax. I hadn't sent or received one of those in ages and I didn't know if the fax machine would work, but I told them to give it a try. A few hours later the phone rang, the answering machine answered, and the fax started coming in. And then it stopped! According to the little square message box there was no paper in the lower tray. But there was!

It told me to put paper in the tray and push o.k. I took out the paper that was already in there, put it back in, pushed o.k. Once again it told me there was no paper in the tray even though there was. I took it out and put it back in several times, pushing o.k. and nothing happened. Then suddenly it told me "paper jammed"! I took the sheets of paper out and tried to look for any jammed paper. Couldn't see a thing! Couldn't get my fingers far enough in to feel anything due to the small space. How was I going to make sure there was something in there? Maybe a mirror that magnified things---didn't work! O.k. what could I use to see in there? I needed a flashlight. Found my light that I keep handy in case the electric goes out. Now, I needed to tilt the fax machine up without upsetting it and shine the light down in that thin space. There it was, a sheet of paper caught way in the back. Now, how was I going to get something back in there to get a hold on it and drag it out? It had to be something thin.

    Had to put the old brain to work. How about the yardstick, it was thin. Maybe I could press the end of it down on the paper and drag it forward. I held up the fax machine on its end, held the flashlight, using the left hand to hold both, while fishing back in there with the yardstick. Didn't work. The flashlight was too big and heavy while trying to hold up the fax machine. Found my pocket size flashlight compliments of the American Dairy Association a few years ago. Tried the procedure again, still didn't work but the small flashlight was easier to hold. Next idea, what if I fastened some tape on the end of the yardstick, sticky side out, pressed it again the jammed paper and pulled --- didn't work! It was too hard to pull out and the tape wouldn't hold. Now what?

    If I could get something under the edge of the paper, maybe I could "scrunch" it up enough to get a hold on it. What do I have that is thin enough to go in and that I could get under the edge of the paper? A fork? Not long enough. Pliers wouldn't work. Think!! Aha, got it! Grandma's long bladed butcher knife that has been laying unused in one of the kitchen drawers for years! Sure enough, I was able to get it down in there, under the edge, scrunch up the sheet of paper, and drag it forward. After a couple tries, I was able to bring it forward enough to reach it with my fingers and pull it out. I put the copy paper back in the tray, pushed the o.k. button, the fax machine started and the copy of the sale ad came through.

So, all it took to get a faxed copy of the ad for the Ohio Holstein Convention Sale was a mirror, a yardstick, tape, two flashlights, my grandma's butcher knife, time and a lot of patience! The moral of the story --- don't throw away all those antique things you seldom use --- you never know when you are going to need them. The electric knife was no help!

I GET BY WITH A LITTLE HELP
FROM MY FRIENDS

I have been involved in so many different activities, experiences and situations in my lifetime and the most important and valuable thing I have gained from all of it is friends. In my involvement with registered Holsteins, the Ohio Holstein Association, farming and so many, many other activities I have met and gotten to know so many people! I am so lucky to be able to call many of them "friend". I look forward each spring and summer to announcing and working with the dairy shows, as I know many of my friends will be there to see and talk with and there will be new ones to meet. We all have the same things in common --- our love of dairy cows and farming.

In times of trouble and tragedy friends are immediately there to give support in any way they can and in any way that it is needed. I definitely know that from experience! I recently read about a young lady who participates in 4-H and her local county fair. She is 14 years old and raises market hogs and her market hog weighing 233 pounds was to be auctioned off at the fair's market sale. She made an announcement that the proceeds from the sale of her hog was to be donated to her friend, who was struggling with pediatric autoimmune neuropsychiatric disorders, known as "PANDAS" and needed treatment that the insurance would not cover. She raised over $10,000 to help her friend. My hat is off to this young lady and to the parents who raised her!

In my area a young boy scout of 15 used his woodworking talent to build benches for the people residing in our local county home to sit on. He worked with the county home superintendent to determine the height and width that the benches needed to be to make them the most comfortable for the residents. I am sure his thoughtfulness gained him many friends among the residents who live there. He will soon be receiving his Eagle Scout award.

I was so glad to read the stories about these outstanding young people. We have so many outstanding and caring people of all ages in our world, yet so little of our news reports and newspaper stories tell us about them. More of this type of news would make our days brighter and better!

For so many, a special or best friend, or "bff" as they refer to them today, is so important. We all need someone we can talk to about our hopes, our dreams, our troubles, our beliefs, our frustrations or anything else and know it will be kept in confidence. I am a firm believer that if more troubled people just had a friend or someone they could sit down and really talk to, knowing that person would just listen and try to understand, it would help to relieve their frustrations and their reactions to the problems would be different and less violent.

There is an Arabian Proverb that says, "A friend is one to whom one may pour out all of the contents of one's heart, chaff and grain together, knowing that the gentlest of hands will take and sift it, keep what is worth keeping, and with a breath of kindness, blow the rest away". In the plan of God, a friendship is a touch of heaven on earth. (Mark Connolly)

THE IMPORTANCE OF THE JUNK DRAWER

Over the years in many homes there has been a special place known as the "junk drawer". I think perhaps it originated back in the depression era when people were so very poor and learned to never throw anything away, as there might come a time when there would be a need or use for it. Large items went to outbuildings or a shed but the very small items were often dropped in a drawer, often located in the kitchen where they could easily be found. A junk drawer has been a part of my kitchen for many years and has often been opened.

On a recent afternoon I went to my mailbox to retrieve an assortment of important and unimportant mail only to discover that there was a large dent in the side of my mailbox, the red flag would not go all the way down, and the rivet in one side of the lid was broken off and the lid was hanging open. In surveying the damage, I ascertained that I needed to find a way to reattach the lid to make it usable until I could get a new mailbox. To do that I would need a very small bolt or screw with a nut that would keep it in place. Where might I find one? Of course, look in the junk drawer. I had not had an occasion to visit the junk drawer in quite some time, so I had definitely lost track of the contents. When you open that drawer you just never know what you will find in there. One thing is sure, some of the contents will bring back memories.

Of course, my drawer was stuffed full, however at some time I had obviously made an effort to organize it. Who knows when I had found time to try to do that! Some things were in plastic bowls, there were different sized jars filled to the brim, there was a recipe box, plastic bags with an assortment of items, and then some things were just dropped in the drawer. I would need to be careful how I put things back or I would not get everything back in!

One of the first things I found lying right on top was my inexpensive cheese and vegetable grater. No wonder I couldn't find it when I was hunting it! I never thought to look in there! There was a recipe box, the size that holds regular recipe cards. It definitely got too small to hold all the recipes I collected over the years. It now holds a new dishwasher snap adapter (I don't have a dishwasher anymore), some brass fittings, there is a new bag of ten screws priced at thirty nine cents (they have certainly been there a few years), a small key on a key ring that looks like an owl (I wonder if that key is for our antique John Deere lawn mower). In later years when he couldn't walk very well, my Dad traveled all over the farm on his little John Deere. The key ring was given by two men, friends for years, who delivered gas and oil to the farm many years ago! There are a couple dominos --- we used to play dominos often and had a lot of fun playing a game called Mexican Train. My dad's Barlow penknife is in there. He always carried it in his left pocket (he was left-handed) and used it for so many things! It cut a lot of baler twine as well as several apples when we would stop to rest the horses under the old apple tree in the fall of the year.

There are three dairy calf weigh tapes. You put the tape around the calf right behind the front legs, pull it tight, read the pounds on it and you have the approximate weight. It was a necessary item for 4-H members in years gone by, as the calf's weight had to be recorded each month in the 4-H project book. Of course, if you needed it at the last minute just before a 4-H meeting and the calf was somewhere out in the pasture, you just added a few pounds to last month's weight. You knew the calf was growing!

There is a glass baby food jar that has definitely been in there for quite a while. It has been several years since there has been a baby in this house to feed --- probably the last one would have been a granddaughter, 30 years ago! Time does get away! It is filled with screws, tacks, small bolts, etc. I probably should throw away most of the things in this junk drawer, but just as sure as I do someone will come along looking for the very thing I threw away! So I will just pack everything back in the drawer--- along with the memories. Oh, I just found the exact size bolt I need and it has a nut on it. Just as soon as I get everything tucked back in the drawer, I'm off to repair the mailbox!

9-1-1 WHAT IS YOUR EMERGENCY?

When you are involved in an emergency situation your immediate thought is to dial 9-1-1. When you hear those words, "9-1-1 what is your emergency?" you feel a sense of relief, as you know help is going to be on the way. In a state of panic, you give your name and address, try to explain the emergency, and try hard to think clearly so you can give the dispatcher the needed information as questions are asked. The voice is calm and re-assuring as you are told to stay on the line, to keep talking and that help is coming. The dispatcher will continue talking to the caller getting information for the responders.

As you wait the minutes seem like hours. Then you hear that first siren! What a wonderful sound! Help will soon be there and they are coming as fast as they can! Often the first to arrive is an officer from the sheriff's department, police department or highway patrol. Accidents aren't always on the roads, they may be in fields, on farms or many other places. The officer will try to locate the emergency, give directions to the responders, handle traffic, make sure no one interferes with the emergency people and help in any way they can. Our deputies and police officers often have no idea what to expect when rushing to an emergency call. There are times when their own lives can be in danger.

The fire departments arrive from different towns in the area. In this rural county the firemen and firewomen are volunteers. They must drop whatever they are doing and rush to their stations to get equipment or go directly to the emergency site. We are so lucky to have these people who care about our community and the people in it. They have given much of their time for special training in all types of situations. These volunteer fire departments need and deserve our support. They are often taken for granted.

The ambulances arrive with emergency technicians and paramedics and they take over. They have attended schooling and been trained to handle many types of emergency calls, although at times they may encounter unusual situations. They must stay calm and try to keep the victim calm, gain as much information as possible and then handle the problems in the best way they can. Their calm re-assuring voices can be so important to those in need as they work to give aid and get the victims on their way to the hospital.

As the word of the emergency or accident goes out help keeps coming. Friends, neighbors and complete strangers arrive to give their help if needed. Not knowing all the details, some bring along equipment just in case there could be a need for it. They do their best to comfort family members and they say prayers. They are there because they care and they will continue to give help and support after the emergency ends.

If you have never had to dial 9-1-1 consider yourself very, very lucky. Those of us who have made the call thank God for it every day and for the help it sends. The people who answer that 9-1-1 call can be the difference between life and death!

REMEMBERING WHEN POT WAS A GOOD WORD

The word "pot" is defined as a container of earthenware, metal, etc. usually round and deep and having a handle or handles, and often used for cooking, serving, or other purposes. When I think of a pot, it brings back many good memories.

When I was young our family usually came together at my Grandma's house for Sunday dinner. As you walked in the door there would be the aroma of all the delicious food awaiting you. On her stove would be a big pot filled with chicken, broth, and her famous homemade noodles, big, wide, golden strips of delight. No one else ever made noodles that tasted quite so good. Perhaps it was that special bit of love that she put into them.

The pot on my Mom's stove was often filled with vegetable soup. Tender chunks of beef, beef broth, onions, corn, celery, cabbage, potatoes, carrots, and tomatoes. All healthful vegetables fresh from the garden, so tasty and so good for you. But never green beans! My mom didn't like them in her soup! However, at other times, there would be a big pot of green beans, ham, and new potatoes simmering on the stove. Add a slice of warm homemade bread and butter --- what a treat.

Often when we stopped to visit special friends, Harvey and Wilma, it smelled like you were entering an Italian restaurant. There would be a big pot of her special spaghetti sauce simmering on the stove. Sitting on the counter would be jars and jars of that beautiful red sauce to be enjoyed throughout the winter months. There was no other sauce that tasted quite like it. Store bought brands couldn't compare.

My grandpa kept one of those old fashioned blue and white

enamel pots on the back of the coal stove. In it was sassafras root covered with some water, always warm and simmering just a bit. My Grandpa made sassafras tea from it and drank a cup every day. He always said it was good for your heart and to thin your blood. I don't know if that was true or not, but he lived to be in his late eighties with no health problems.

I find it sad that the word "pot" has a very different meaning today and so often is part of the headlines in the newspapers. I wondered how in this day and age the word for a perfectly harmless container came to mean marijuana and be associated with narcotics. I discovered that "potacion de guaya" is the name for wine or brandy in which marijuana buds have been steeped. Evidently the news media shortened those words to "pot" when writing about marijuana. It was probably much easier for them to spell "pot"! Not everyone excels in spelling!

I am certain that when Herbert Hoover spoke about "a chicken in every pot" he never thought it could come to mean "free range hens in a marijuana patch"! The word "pot" today often applies to very unpleasant circumstances. Too bad that is what the people today will remember. Words should bring back pleasant memories.

THE IMPORTANT SIGNS OF THE TIMES

Have you ever stopped to think about how many different signs we encounter as we go through our lives each day? There are many different types of signs and they are very important in many different ways. Of course, the obvious are the road signs. We rely on them to guide us and help us reach our destination, whatever that might be. We need them to show us where people live, where businesses are located and more. Without them we would become lost. Think about the early settlers in our country. The only thing they had to guide them was the moon, the sun and the stars. As they traveled the country they would make a camp along the way and then go out to hunt for food or scout the area. They had to rely on rocks, bushes, trees and streams to help them to find their way back to their camp. Once a settlement was established signs were made to help guide people as they traveled.

In those days, as buildings were put up for businesses, signs were put on them often with just the owner's name. As the years went on many of those names would continue and some of them exist even today. When we would see them we usually knew what items could be found in those stores. Smaller business generally put up signs telling what items they sold or the service they provided. When something was needed you knew just where to find it. Just imagine if there were no signs in our world today!

The early settlers didn't have the seasons marked on a calendar to tell them when they would arrive as we do. They depended on the signs of Mother Nature to let them know when to start preparing for the changing season. People often try to get a "jump" on Mother Nature, especially in the spring, but she usually has the last word on when the season can get going and crops and plants can grow. For many years there have been those words --- "don't try to fool Mother Nature"!

There are also many signs in life that we can't see. There are signs that have to do with health and well-being. There are many of those that we feel that warn us that we are coming down with a cold, aches that let us know there is a problem with our bodies, etc. These signs should not be ignored, but people often do not pay attention to them and then must deal with serious consequences. We are so lucky to have the doctors, hospitals and equipment to diagnose the problems and take care of them.

There are also signs that can never be explained. Often in life there are problems that occur for which there are no obvious answers. We think about a problem, mull it over in our mind, spend sleepless nights and just can't come up with an answer. So you say a prayer and ask for a "sign" that will help us to make the right choice or wise decision. You never know just what that sign might be or how it might appear and it often comes to you in an unusual way. You don't question it, you just accept it. This is one of those days when I am looking for a sign! Keep your eyes open for the signs to keep you on the best path in your life.

DON'T FORGET THE ROLLED CRAPPIN' PAPER

We are being warned about a week of upcoming severe winter weather. There is to be more snow, below zero temperatures and minus wind chills. Everyone needs to prepare for the weather conditions ahead. Roads could be impassable and there is always the danger of losing the electricity and sources of heat.

One of the first things that is important is water. A supply of bottled water should be laid in or clean jugs filled and set aside so there is water to drink and cook with in case pipes freeze. Forget about baths for a few days, just have plenty of deodorant and perfume available.

Should the electric go off, you will need a supply of milk so you can eat breakfast cereal. Bread will be needed and having some ham, baloney and cheese on hand would be a good idea. No one is going to starve when there is peanut butter and jelly around! Canned fruit is good and there are worse things than cold baked beans. Be sure to have one of those hand-held can openers --- the kind that are so hard for old people with arthritis to use. The older people may need to stock up on those high protein drinks so they have extra energy for keeping warm while they nap.

Be sure to have flashlights and extra batteries. Candles are nice, a lot of them are made to smell good these days and just sitting around watching them flicker is just about as enjoyable as watching most of the television shows available these days, however they can be dangerous.

Put on your warmest clothes and lay out extra blankets in case your heat goes off. If you can't take a bath you can just keep wearing the same ones every day and you won't lose any body heat by changing. Everyone will be so concerned about surviving they aren't going to pay any attention to whether you have on a matching outfit. Company is unlikely anyway!

And one of the most important things to remember ---make sure you stock up on rolls of crappin' paper! In the "good ole days" there were different names for the outside bathroom facilities. It was called a "privy", "outhouse", "loo" but cowboys often referred to it as the "crapper" (I watch a lot of old western movies). Back then the "crappin' paper" was very important. Probably not too many of you reading this remember how important it was to the crapper user when the Sears & Roebuck catalog came out in the 1800's! The catalog was often kept hanging on a nail. Its use was very popular until they started using glossy paper in the 1930's.

Toilet paper was first introduced in 1857. Joseph Gayetty is credited with being the inventor of the modern commercially available toilet paper in the USA. Gayetty's Medicated Paper was sold in packages of flat sheets, watermarked with the inventor's name. The ancient Greeks used flat stones and pieces of clay to cleanse themselves. The ancient Romans used sponges on sticks, which were kept in buckets of water for everyone to use! Corncobs were used in colonial America and later random paper products were used. Americans use seven billion rolls of toilet paper a year. Some countries do not use toilet paper. They have hand held showers or bidets.

It was also important years ago to know what to do if you were working out in the fields and "nature called". You didn't have paper towels or tissues in your pockets then. When you picked your spot, you wanted to be absolutely sure there was no poison ivy in close proximity. Getting too close to it could bring about a total disaster! You then needed to look for "plantain leaves", as they were safe and fairly comfortable to use in such emergencies.

My cupboard and freezer are well stocked. No matter how deep the snow gets or how long it takes to get the roads open, I don't have to worry. I learned months ago that my cousin's church collects and recycles paper, and I started saving paper for him, including all those letters wanting money that I have received recently! However, he is a busy fellow and has not had time to come and collect my bags filled with paper, so I have several stacks in my basement. Thus, if there is an emergency, my supply of "crappin' paper" is unlimited! Let it snow!!!

DEALS, DISCOUNTS, FREEBIES AND FLIMFLAMS

One of the joys of being retired is the privilege of being able to sleep in late in the mornings. For years the alarm went off early and it was roll out, get dressed, grab some coffee and head to the barn to milk the cows and do chores. How good it feels since retirement to pull those warm covers closer on a cold winter morning and stay in bed. So much for that idea! It isn't the cows that cause you to wake up early these days, it's the telemarketers and robocalls!

They didn't even wait until I had recovered from my New Year's Eve celebration (did the ball drop? I fell asleep). Early on the morning of January 2, 2019 I was awakened by the ringing of the telephone and it was a telemarketer! The calls have been almost non-stop, four or five a day, with all kinds of offers. I can get a free device to alert someone if I need help, of course to keep it I must pay a fee every month. The warranty is about to run out on my car and for a fee I can extend it --- on a 1981 Ford Ltd that has met up with deer twice! There is the caller who says, "Don't hang up, this a survey and we need your opinion (I am good at giving those), however in order to keep conducting these surveys that influence the politicians we need your support. Can you send us some money?" (The only thing that influences politicians IS money --- not surveys!) There are calls about the electric company that doesn't provide my electric, calls about credit cards that I don't have, it goes on and on! It is especially irritating when then want to talk to my husband who has been dead for over five years!!

And then there are the letters in the mail. The New Year has brought on letters daily wanting money for something.

They send me things I didn't order or ask for and then want me to send money for them. I have so many labels with my name on them, I could practically wall paper a room! The number of magazines that have sent letters to me wanting me to subscribe is unreal. If I subscribe at the reduced rate they are giving me I will receive not only the magazine but also free gifts. I love to read and I do read about many different subjects. I enjoy getting a magazine and sitting down with a cup of coffee or tea to look through it. However in so many of today's magazines there is little of interest to read. When you open it and start to look at its contents you find page after page of advertisements before you find anything to actually read! One of the magazines offered a subscription for only $20 and they were giving me a discount of $99.80, a real savings. Their magazine would tell me all about the "important people", movie stars, television personalities, the " royals in England" --- important subjects!

There was the one all about living in another part of the country where there are "picturesque parks", "relaxing getaways" and "dynamic districts". We have two hotels, Atwood Lake, Bluebird Park, the Algonquin Mill and buffalo in this area! Once again I would be getting a great discount and free gifts --- "A Slow Cooker Cookbook", "Pin-Worthy Recipes Cookbook" and a "Best Comfort Cookbook". Now what in the world do I need with all these cookbooks? Not only am I all alone, don't eat very much, don't do any more cooking than I have to and am supposed to be on a diet to lose five pounds! Besides, if I am lucky enough to be going somewhere I am going to eat at a local restaurant before heading home!

Most of the letters did not reveal how much the postage cost is to send them out, however one did and it was only 21 cents, so I have to assume the publishers get a break on postage cost. When I mail a letter my cost is fifty cents and in a few days postage goes up to fifty-five cents. Perhaps, if all the items and letters that are nothing but a nuisance to us were to cost the sender the same as we pay, we wouldn't receive so many! Thank goodness for farm magazines, dairy magazines and dairy breed magazines that give us interesting articles and interesting people to read about and learn about! If I could get an answer when trying to call those numbers that come up on the phone with telemarketers and the robocalls, those responsible would have an opportunity to learn what it feels like to get up to milk cows at five o'clock in the morning!

WEDNESDAYS ARE FOR SHOPPING WITH EMILY

Another year --- another birthday! Each time I add another year to my age, I try to understand some of the changes that happen to me. Of course there are some that I just have to accept, as there is no point in questioning them. Grey hair --- I seem to notice a little more each time I comb my hair. I always did want to be a blonde, but coloring is not for me. A shorter haircut will eliminate some of it --- I must call for an appointment! Lack of energy --- I used to rise early and be on my feet for hours, milking cows, running after children, cleaning house, cooking for the family, running to town for parts, etc. Now making coffee, getting food out of the refrigerator, putting it in the microwave and pushing buttons wears me out! And speaking of lunch --- I used to feel hungry and required a large portion of food at each meal. These days food seems to be very unimportant to me and I eat smaller amounts. Maybe that is because these days I am "required" to follow a "healthful diet" instead of being allowed to eat all the things I learned to love over the years. Mashed potatoes and gravy and pecan pie are much tastier than kale and broccoli!

Why do I put off doing house work so often? While I never really liked doing it, I knew it had to be done, so I did it. My mother was a very neat housekeeper and I always thought she would be checking on me. I still make my bed as soon as I get up every morning! I am always "postponing" things that need done in the house. For instance, today I got the sweeper out of the closet and there it sits in the kitchen. That is as far as I got with it. I had an idea for a column and I had to write it down. Maybe tomorrow I will get the sweeper plugged in!

Shopping --- I used to enjoy getting away to go shopping. A friend and I would go to the mall and walk around and look at all the things we had no money to buy. And we always stopped for lunch somewhere to talk. With taking care of the cows and everything else I had to do, I seldom had time to go. Now I could go every day if I chose to, but I have absolutely no interest! My house is comfortable and has everything I need. My closet and drawers are full of clothes and I usually wear the same few outfits most of the time. And I certainly don't want anything more to clean or dust!

However, I did go Christmas shopping with my granddaughter, Kristin, her three year old daughter, Emily, and her one year old son, Daxton. Kristin called in the evening to ask me if I wanted to ride along the next day as she needed to pick up some things and do some Christmas shopping. It is a help to her to have someone along to sit in the car with the children if they are sleeping or when she just needs to run into a store for a quick errand. I had nothing of any importance planned as usual, so she picked me up the next morning and away we went.

Following the quick errands, it was time for shopping in the big stores, something I had not done in quite a while. We each took a child in a shopping cart and away we went. Kristin led and I followed, trying not to get lost in the crowd! The children were both very good (I am their great-grandmother, would I tell you anything different)! I could hardly believe the size of the first store and the huge amount of items in stock! When Kristin got done her cart was full, including chicken feed and a salt block, I still had just Daxton in mine and sticker shock! Then it was on to the next big store for groceries and much more walking than I am used to!

Emily is a smart little girl for three years old and she talks a lot. She has dogs, Lily is her dog, a cat, fish and chickens. According to Emily, they lay eggs that you have to get, they run around the yard and sometimes go across the road, and when they fly up on the porch railing you have to shoo them off! Three of the chickens have names --- Chickaletta, Henrietta, and Bernise. Bernise?? This is only a small portion of the important knowledge that Emily shared with me on our shopping trip! It was such a joy to spend time with my granddaughter and her children!

While age can change your ways and create problems that you have to deal with, it also gives you the time and opportunity to enjoy life's greatest gift --- children --- grandchildren and great-grandchildren. As we were eating dinner at the table on Christmas Day I asked Emily when we would be going shopping again. There was a moment's hesitation and then she replied, "Wednesday, we will go in Mom's car and I'm driving"! That's Emily! I made a notation in my datebook that I hope will happen for years ---"Wednesday --- shopping with Emily".

CAN SHE MAKE A CHERRY PIE, BILLY BOY, BILLY BOY

That is a line from an old folk tune written years ago. It pointed out the importance of asking questions before selecting a mate. Cooking and baking are creative arts. It is a talent that needs to be used and exercised on a regular basis. If not, your "artistic talent" and your ability lessens. And so it was with me. I decided one day last week to bake a couple pies. It had been over four years since I had baked my last pie – a cherry one for a special friend.

In his later years my husband had diabetes and had to be very careful about his diet. He didn't have the needed willpower to avoid baked goods and his favorite was pie, especially black raspberry. We didn't care for using so much of the artificial sweeteners, so I just stopped baking. If the baked goods were not available it was much easier for him, and having been a chubby soul all my life, I didn't need it.

My Mom's family has always had its share of good cooks. Of course many had foods they were known for, Aunt Ruth was chocolate cake, Grandma was fried chicken and homemade noodles, Cousin Polly was graham cracker crème pie and the list went on. My Mom was an excellent cook and was known for many things, including her raisin filled cookies. There also was Aunt Gertrude who was known for her homemade dinner rolls. She brought them to every holiday meal or picnic. They were terrible! But no one would ever hurt her feelings by telling her. They were always placed in a nice basket on the table with the rest of the food. When the meal was over the basket would be empty. Where Aunt Gertrude's rolls disappeared to was always a well-guarded family secret!

I had a family to cook for, a husband who loved to eat and three children, including two sons who were always hungry and grew up to be well over six feet tall. With my Mom's help and family recipes, I learned to be a pretty good cook and baker. When I baked pies it was usually six or eight at a time. However, I learned the other day that to be a great "artist" of any kind requires practice on a timely basis. My pies were o. k. but I have lost a little of my "special touch".

We are so lucky to have fresh vegetables from the gardens, all kinds of fruits, fresh meats, dairy products and so many things available for cooking. When one cooks every day for a family you develop a special ability and talent just as any painter, sculptor, musician or other talented person. That old saying "practice makes perfect" applies to cooks and bakers also. In this day and age so many families rely on frozen dinners, take out, or fast food because of their hectic schedules. They just don't have the time to practice the art of cooking and baking. But for those of you who still cook and bake, the food you prepare and set out on the table every day is a special type of "art", so important to the health and well- being of your family. Take pride in your creations!

GRANDPARENTS ARE SPECIAL PEOPLE WHO DO SPECIAL THINGS

My house currently smells like an Italian restaurant! At the request of my granddaughter, I have been making and canning spaghetti sauce. Since the dry weather and groundhogs ruined our garden, it has been necessary to purchase the tomatoes, onions, and green peppers, as well as the oil, sugar, oregano, and black pepper. I had a box of salt! For what all of this is costing, I could have purchased twice as many jars from the store, but that wouldn't be "Grandma's spaghetti sauce".

Grandparents are special people who do special things. I remember how great it was to be allowed to spend a few days with my grandpa and grandma, who lived in a small town, not in the country. My grandma made the best homemade noodles ever! Grandpa always kept a special bag of candy tucked away in the back of the cupboard just for us grandkids. He would walk us up to the local grocery store for ice cream. Going to bed at night and sinking down into that warm soft "feathertick" on the bed at their house was a special experience. I was afraid I was going to sink too low and wouldn't be able to breath!

Our daughter had beautiful long hair that she wore in braids for several years, but my mother, her grandmother, was the only

"Biddy Buddy & Cindy"

one who could braid her hair. She didn't like the way I did it, so every morning before she went to school, she ran across the driveway to Grandma's house to get her hair braided. And my mom was always waiting to do it, even though it meant getting up early every morning. In later years, when her three daughters, my grandchildren, stayed overnight, they went every morning to "Biddy Buddy's " (their name for my mom, anyone remember George Gobel?) house so she could make them "dippy eggs" for breakfast. No one else could make dippy eggs like hers!

I never knew my grandmother on my father's side, as she had passed away when my dad was two years old. My grandfather lived with us for awhile, and he sat me on his lap and taught me how to read and enjoy books when I was four years old. I had broken a wrist and couldn't play very well, so we spent a lot of time together. No doubt he was a great influence on my love of reading, my abilities in school, and my writing.

For my grandsons, who are both attending the University of Tampa, my specialty is frosted sugar cookies, and I will be looking forward to greeting them with a supply when they come home for the holidays.

Grandmothers and grandfathers are very special people to children. They often bring special "gifts" into their lives that aren't always recognized until children are grown up. For many, the grandchildren live far away and can only visit once in awhile, but we are fortunate today to have not only telephones but computers with skype so we can see one another and visit together. And if all else fails, there is still the postal service and letters that can be written. No matter how it is done, staying involved in grandchildren's lives is very important and special to all concerned. We are so blessed to have them! And if you don't have grandchildren around you, there is always a child who needs an extra hug or some special attention. There are times when just having someone to talk to can make a difference in their lives. They don't always have to be "your" grandchildren!

Jan 1982
George and Laura Wagner - 50th Wedding Anniversary

BAGGY SOCKS JUST WON'T DO

The weather is cold and there is snow on the ground. No matter where we go or what we do, the important thing is to bundle up from head to toe and stay warm. One of the most important items in our wardrobe is socks to keep our feet good and warm. The first known socks were animal skins gathered up and tied around the ankles. The Greeks had what they called "piloi" which were made from matted animal hair. Eventually the knitting of socks became popular and many evenings were spent by the ladies knitting socks in preparation for the coming cold weather. Then the making of socks by machine came about.

Most children like to go barefoot in the summer or where the climate is warm year round. As a child I could not go barefoot very often. My grandpa lived with us and he had several hives of bees that he enjoyed tending. Our yard was covered with tiny white clover that his bees loved. If I went out in the yard in my bare feet I was sure to get stung! So I learned at an early age to wear my socks and shoes. It became such a habit as I grew up that I seldom go barefoot even in the house. Growing up there was a neighboring family with three boys my age and as soon as the first warm spring days came they were in their bare feet. I used to marvel at how they could run through the stubble in the wheat fields, down the slag roads, through the pastures and over the hills without ever showing that anything hurt their feet. The soles of their feet must have been like leather!

I like socks that fit my feet snug and tight. Baggy socks just won't do! I don't want my socks bagging down around my ankles or all stretched out! In my high school days "bobby socks" were the thing. They had strong elastic tops that you could wear straight up or fold down to the length you wanted. They were well made and long wearing. Today it is hard to find socks with elastic that will last through very many washings. There are many types of socks and stockings available these days for both men and women. They come in many different lengths for all types of shoes and boots. There are many colors and designs, with many featuring pictures or sayings. The choices are limitless!

As we listened to the tributes to and heard the stories about President George Herbert Walker Bush recently, we heard about the socks he and his family wore as a tribute to First Lady Barbara Bush at her funeral. John Cronin is the founder of John's Crazy Socks. He is a 22 year old with Downs Syndrome. He designs and sells socks and is very successful. He had heard about President Clinton sending a gift of socks to President Bush, so he decided to send a box of his socks to him. In return he received a very nice letter. He designed book themed Downs Syndrome Super Hero socks and sent to him. A percentage from the sale of his Library Socks for Literacy is donated to Barbara Bush's Literacy Foundation. John has designed President George H. W. Bush tribute socks and they have completely sold out. John's success not only shows what someone with Downs Syndrome can accomplish, John also spreads happiness through his socks.

At this time of year another type of sock also becomes very important --- the Christmas stocking. There are no written records of the origin of the Christmas stocking. There are only popular legends that attempt to tell the history of the Christmas tradition. Children all over the world hang their stockings in anticipation of Santa coming to fill them and hopefully leave some presents under the tree. They are warned to be good during the days before Christmas or he might just leave a lump of coal in their stocking! While socks may seem to be a simple part of our daily life, they can be very important not only at Christmas time but especially if you have cold feet!

EXPERIENCE THE JOY OF
SLEEPING IN STRAW

Many of us living and working on dairy farms have slept in straw. It might have been an all-night vigil waiting on the birth of a baby calf or staying up late with a very sick cow. It could have been when working with a show string, preparing the space to tie them in, trucking the cattle, washing and clipping, milking the cows, getting everyone fed and settled, and walking the line to keep every animal clean. When the chance came to throw down your sleeping bag on some hay or straw, it felt just as good as the memory foam in an Ikea mattress in a plush hotel! Or it might have been just needing a few minutes of rest after unloading several wagon loads of baled hay or straw. There are many people who have never known the pleasure that can come from the chance to "sleep in straw".

In the summer time the dairy cows in parts of Switzerland are moved to the high meadows of the Swiss Alps. This leaves the dairy barns empty. Some of the people then participate in Switzerland's "Schlaf-im-Strol" --- "Sleep in Straw" program, which is designed to bolster the income of Swiss farmers and delight visitors from near and far with a unique agritourism encounter. With a thorough cleaning and fresh straw, barn stalls make affordable accommodations for travelers. Some of the farms offer vacation apartments or chalets for overnight visitors. Barns are supplied with fresh hay or straw, sleeping bags, and blankets. A small kitchenette and rest room is built into a corner of the barn. Guests can choose to stay one night or as long as they would like. Some farms serve a typical "Bauernfruhstuck" – farmer's breakfast. Guests can gather the eggs for that breakfast, help with feeding other animals, or with any other morning chores. They can hike, bike, or ride horses. Farms vary with what they have to offer. They can continue on their way or take time to relax and enjoy the sights and sounds of the country. There are annual inspections to ensure safety and sanitation and co-ordination of marketing efforts by the Swiss Farmers Union. Some of the people who are most interested are hikers, bikers and outdoor enthusiasts. Many tourists are eager for economical accommodations where they can throw down their sleeping bags, stay the night, have breakfast and continue on their trip. The idea is not only popular with tourists but also with local people. The farmers like the idea of improving the image of agriculture as well as the extra income.

Our farmers today not only need ways to educate consumers on farm life, they also need ways to increase their income. Could a "Sleep in Straw" program work in our country? Many farmers have sold their dairy cows or livestock and barns are empty. There are different ideas that would work for any type of farmer. Some of the dairy farmers who have sold their cows are expert showmen. Why not create a program and invite 4-H and F.F.A. members or junior members of breed associations to come "Sleep in Straw" and learn how to clip, show, and judge dairy cattle. Farmers with other types of livestock could offer similar programs. There could be many interesting ideas for youth groups.

Remember how excited you were when you were young and were allowed to sleep over night at the county fair? Both youth and adults from the city would be interested in the opportunity to spend a night in a barn and "sleep in straw"! They could participate in chores around the farm and learn how farmers do things. They could learn how vegetables are grown and harvested, gather pumpkins in the fall, bottle feed baby calves, gather eggs, feed other animals and many more things that are a part of farm life every day. Families could experience those activities, learn the sights and sounds of farm life and just enjoy the time spent together. Our farmers today need ways to educate consumers about farm products and farm life and they need a way to generate more income. Inviting people to "Sleep in Straw" might be an answer. The idea seems to be successful for farmers in Switzerland.

FARMERS HAVE FELT THE WINDS OF CHANGE THROUGHOUT THEIR LIVES

I recently read an article in a local farm paper about the research that is being done for new housing for chickens. Consumers of eggs and chicken in this day and age want to see labeling with words such as "cage-free", "natural" or "free range". As I read about the research that is being done, my thoughts turned to years ago and my Mom and her chickens.

There was always a chicken house and chickens and my Mom enjoyed working with them. The chicken house was a wooden building set upon cut sandstone. It had a dirt floor, a high roost for the chickens to rest on at night, wooden egg boxes filled with straw along the walls for the hens to lay their eggs in, a long feed trough for them to eat the chicken feed  out of with a roller on top so they could not fly up and sit over the feed, and large metal water containers. There were windows but only one door that was closed every night to keep the chickens safe.

Early every morning when mom went to the "outside" toilet (we had no inside one) she opened the door so the chickens could come outside and roam. I guess now-a-days we would call that "cage free". As the hens came out they would immediately start picking bugs out of the grass and head out to look for the many other things they would scratch up and eat during the day including the piles of cow manure in a nearby pasture. They would travel all over (free range) the field, barnyard, lawn, etc., eating whatever they chose. Although at times they were close to the road, they never seemed to go there, and I wondered how they knew to avoid it. Those chickens were very important to mom. Some were pets and had names. Not only did they mean fresh eggs for breakfast and to cook with and chicken for Sunday dinner, they also provided a few dollars from selling the extra eggs. That could mean a new dress or pair of shoes, new curtains for a window, or a little bag of penny candy when we went to the grocery store.

There was always a big old rooster with the hens and I was always afraid of him, as he liked to "flop" everyone and he had those big "spurs" on his legs. One day when I went out to play he came running at me, flopping his wings and looking mean and ferocious. I started to scream and my faithful shepherd dog, Stubby, came rushing off the porch, ran over that old rooster barking and biting at him, feathers flying everywhere. When they finally got gathered up, Stubby chased him all the way to the chicken house. That old rooster never bothered me again --- Stubby was my hero!

Collecting the eggs from the nests was fun, except for the time when upon reaching into the nest the hand found a huge black snake curled up in the box instead of the eggs! That old saying "look before you leap" became "look before you put your hand in the hen's nest". A friend of mine told me about the time one of her family members found a blacksnake in their chicken house and upon killing it discovered that it had swallowed a wooden "nest egg"!

The chickens were always free to roam, however there was a time when their freedom was limited. We had a registered Border Collie pup and as he grew and started herding, he was very smart and wanted to be busy. When he didn't have anything else to do he would go round up all the hens and herd them into the chicken house. When every last one was in, he would lay down in front of the door and keep them there. Mom would have to go get him and keep him in the house so the chickens could roam.

There is a story that has been handed down in my family about a special hen that my great-grandmother had. She delivered her eggs! In those days there was an open porch on the house and on that porch was chairs and a rocking chair with a cushion in it. Each day the hen would come to the porch, get up on the cushion, lay her egg and then leave!

Over the years there have been so many changes in the way all farm animals are raised and cared for and there will be more to come. Every farmer, regardless of the type of livestock he raises, wants to do a good job of feeding and caring for them, as his goal is to make a living. Farmers are dependent on the consumers to purchase his product or foods manufactured using his products. Therefore the consumer's concerns about how livestock is raised and handled must be considered. We need the consumer to buy and use our products and we must do whatever it takes to achieve that. Accepting change isn't always easy, and often we don't agree with the change, but there are times when we must accept it to survive. William Arthur Ward wrote: The pessimist complains about the wind; the optimist expects it to change; the realist adjusts the sails.

OLD HABITS REALLY DO DIE HARD

I highly recommend that after you reach a certain age you stop moving things around. Leave everything exactly where you have kept it for years. I guarantee that if you move something you will then have trouble finding it no matter what it is! Over the years we tend to keep items in the same places until one day, for some unknown reason, we think we should change the location. Mistake!

One day last week I was on a "mission" to clean my kitchen and put up new curtains. Since I am at that age where I should not be climbing a step ladder, a family member came to help me. In the course of cleaning, a piece on my microwave cabinet was broken. Both the microwave and the cabinet have been around for a long time. There are small cabinet doors on it and the latch on them has been broken for quite a while so I have to set something again the doors to keep them shut, and a piece of trim on the bottom is missing. Even though the family member was upset at what happened, it wasn't a "big deal", as I need to get something new.

I have another cabinet in my kitchen and she suggested moving the microwave to it. I agreed that might be a good idea, however the only place to plug it in is behind my refrigerator, too far away for the microwave cord. Her next question was, "What about getting a power strip?" Hmmm- good idea! The next day I mentioned the idea to my son, who was planning to come to visit. When he came on Saturday, he brought a power strip, attached it to the cabinet, hooked things up and moved the microwave. It looks good in its new spot and I like the location. Just one problem --- I keep forgetting where it is!

The microwave has always sat on the cart in a corner of the kitchen right behind my seat at the kitchen table. I could put something in to warm and when the buzzer sounded just turn around and retrieve it without getting out of my chair. It has only been a few days since I moved it and I have already made numerous trips to that corner with whatever I wanted to warm in spite of the fact that the microwave is in plain sight in its new spot!

My daughter-in-law suggested that it would probably be safer for me in the new location, as there is limited space getting around the table to the old location and I keep a throw rug there that I could trip over. Since I have to get up from the table and walk to the microwave to get the food and back to the table, I will be getting more exercise. Everyone – doctors, family, etc. are always telling us "old folks" to exercise more! Most of us feel we have had enough years of "exercise"! We just want to rest!

When you have kept things in the same place for years, it is very hard for the mind to accept the change. Your mind has developed the habit of going to the same place each time and the habit is very hard to break. There are so many farmers and others, who have always awakened at an early hour in the morning, who never needed alarm clocks. After retirement, when they could sleep in, they still awaken at that early hour. When the brain has been programed a certain way for so many years changing it is very difficult.

I remember my dad telling a story about a friend of his when he was young who would stay out later at night than he was supposed to. So when he went home, not wanting his parents to hear him, he would take off his shoes, tiptoe to his room, shed his clothes and jump into bed (one of those old-fashioned ones that sat high off the floor), leaving all the lights off. One night he went home late, followed his routine as usual and jumped. Just one problem, his mother had cleaned his bedroom that day and moved the bed! So remember---keep everything in the same place, unless you want to get more exercise while looking for it, and make sure you know where your bed is! Old habits really do die hard!

HAVE YOU GIVEN ANY THOUGHT
TO THE HEREAFTER

The question was asked, "Have you given any thought to the hereafter". The reply was "Oh, yes, I do it all the time. Wherever I am ---upstairs, in the kitchen, or in the basement, I ask myself, "Now what am I here after?" As I stand with the refrigerator door open looking at the milk jug, the cheese, the carrots, the catsup bottle, I keep asking myself, "Now what am I here after?" And then I finally remember, the butter for my toast that is getting very cold as it waits!

There I am in the grocery store without my list, which is laying at home on the kitchen table. As I wander up and down the aisles asking myself "Now what am I here after?", I keep dropping items in my cart, just in case they were on my list that I can't remember. Pretty soon there I am pushing a cart out the door, full of groceries that cost almost $100 and I just hope in one of those bags is "what I am here after"!

As we age, memory problems arise and it can be very frustrating, but I find things about the brain fascinating. In school we had to memorize and learn so many things. Our brains were filled with history, mathematics, English, science, algebra, languages, and so much more. These days I can't help but wonder if it was really so important. Why do I remember "In fourteen hundred ninety two Columbus sailed the ocean blue" but I don't remember which ocean or what he was looking for! Why did that rhyme stay in my brain but not the other details? When Columbus and his crew finally found land, did he look around and ask, "Now what am I here after"?

There are many memories from my childhood but only certain ones. Am I really remembering all of those things that happened when I was small or am I remembering things that were told to me. I remember being in the doctor's office when I broke my wrist and being told to count as I went to sleep and refusing to do it. I remember sitting on my Grandpa's lap and learning to read because I couldn't play very well with my broken wrist. I was four years old. I remember many years when we went to my Grandpa and Grandma Gallon's for the holidays. There was a huge family get-together, my Grandma's homemade noodles, and good times together. Aunt Georgia was always late getting there! There are so many things to remember about family, friends, and neighbors as we go through life. I may not think about someone for a long period of time and then a name or word will bring memories flooding back. Where have they been all that time? And it is so frustrating when we meet and start talking to someone we know well---we just can't remember their name! It happens to all of us.

Over the years those of us who are farmers have so many animals in our lives, horses, dogs, cats, cows, and many more. There are so many memories created by our work with those animals. We have so many different reasons for remembering them! For me there are so many memories connected to my Registered Holsteins---their personalities, the things I learned in breeding and working with them, the shows we participated in, the special things we accomplished with them, and especially all the people we met along the way because of them. Those are very special memories.

Why do we remember bits and pieces of all those things we learned in school or things that happened to us in life? Why is it that a certain picture, word, or name triggers memories that we hadn't thought about in years? Our heads aren't very big, so how is all that "stuff" stored in there? There are a lot of questions. Hopefully someday there will be answers. As we age the "Now what am I here after" seems to occur more often. There is no doubt that in the future I will find myself standing in a room asking, "Now what am I here after". When I finally remember what it is, the next question will be, "Now where did I put it?"!

ALWAYS TAKE TIME TO
READ THE INSTRUCTION MANUAL

I am a very lucky person and well taken care of by my family. Often when I need something, a brown truck pulls into the driveway and magically a box or package appears. And so it was a few days ago when UPS brought me a box. My need --- a new phone --- and when I opened the box there they were --- four phones and the base. Having had some health problems, my son is making sure that I have a phone close by at all times.

I started unwrapping pieces, laid them all out on the kitchen table and prepared to try to put everything together. Then I proceeded to read the instruction manual. First was a page of "warnings" and "caution" to "prevent severe injury and loss of property or life". I didn't realize a telephone could be so dangerous! One of the warnings told me to "unplug the product from the power outlets if it emits smoke". There are times when a special friend and I talk so long that I wouldn't be surprised to see the phone smokin'! We have a lot of things to talk about and many world problems that need solved.

"Caution – never install telephone wiring during an electrical storm". As if I am calmly going to be installing telephone wiring when lightning is flashing everywhere, thunder is booming, wind is blowing, and rain and hail are pouring down. I will probably be headed for the basement! And, of course, don't talk on the phone while in the bathtub. They failed to tell me that I might talk too long and the water could get cold.

Next was the "setting up" and all those interesting pictures. Why do we have to have all those confusing little pictures that no one can figure out? The only one that made any sense to me was the one that looked like a clock for setting the time. One was an open ended wrench which indicated to me that a bolt or something needed tightening instead of "initial setting"! One of those notes told me not to plug the AC adapter into a ceiling mounted outlet. How many people have plug- in outlets in their ceiling? I live in an older home and I am lucky to have a plug- in outlet in each room. When my house was built they used oil lamps! Up next was the identification of the "controls". There were pictures, symbols, and numbers to identify everything and to identify display items on the handset. One of those was a picture of a phone that will be displayed to tell me the phone is in use and it will say "line in use". If I have the phone in my hand and am talking on it, I think I will know I am using it!

There were three pages entitled "Making and Answering Calls". One of the instructions was "You can answer calls simply by lifting the handset off the charger". I guess they are assuming that if the phone rings I won't know enough to answer it! Under making calls using the base unit, the first instruction was "dial the phone number after lifting the corded handset" --- in plainer words—pick up the phone and dial! When finished talking, place the corded handset in the cradle". In other words--- hang up! Do they think aliens are purchasing these phones? There are all those special features --- Hold; Mute; Flash; Shared Phonebook; Programming; Special Programming; Caller ID Service; Answering System; Voice Mail Service; Intercom; Transferring Calls; Error Messages and Troubleshooting. Modern technology! All of this certainly makes you long for the "good ole days" when you cranked up the wall phone, gave the operator a name or number, and talked to the party you were calling, while all the neighbors on the party line listened in!

GOOD BROWNIES! GOOD FRIENDS! GOOD MEMORIES!

As my granddaughter and I recently welcomed consignors and organized the paper work for a recent dairy sale, a good friend stopped by to say "hello" and brought us a plate of brownies. So chocolaty---so filled with nuts---so delicious! So nice of her! They are always so perfect with a big glass of cold milk!

The first person to put a recipe for "brownies" in a cook book was Fanny Farmer, who adapted her cookie recipe to be baked in a rectangular pan in the 1896 edition of The Boston Cooking School Cookbook. However, that recipe contained no chocolate! Farmer had made what we today call a blondie. In the late 1890's two advertisements referring to brownies appeared. The first, in the 1897 Sears Roebuck catalog, advertised brownies underneath the heading "Fancy Crackers, Discuits (sic), Etc", but those treats could have been either chocolate or molasses based. In 1906, Fannie Farmer published an updated version of her cookbook that included a blondie recipe and a brownie recipe, both called brownies. After that the recipe started spreading nationally.

I remember so well the first time I tasted brownies. It was at a Farm Bureau Council meeting that I attended with my Dad and Mom. I don't remember the year but do remember it was in the fall. The meeting was hosted by Guy and Dorothy Stine. They were hard working dairy farmers with a well-kept farm, a good herd of cows, and a big, old house with nice things in it including a piano. As the meetings always went, there was a business meeting, a discussion on a subject usually provided by Farm Bureau and then there was food and fellowship. Dorothy was an excellent cook and there were hot sandwiches, salad, home canned pickles and relishes, and then dessert ---brownies --- a new recipe she had found. The brownies brought about a lot of discussion among the ladies present, as no one had made them before. There were many requests for the recipe. I thought they were wonderful, so chocolaty and full of nuts, just like the ones my friend brought us the other day!

As I thought about that time, so many years ago, I also thought about the people there and their lives. There were the three Fisher brothers and their wives. They milked cows, raised sheep, raised hogs, did custom harvesting for other farmers--- ran a grain binder, baled hay and straw and took their threshing machine all over the area. Their wives and children usually milked the cows and took care of the livestock. The Shawver brothers also did custom work, raised sheep and were known as the best sheep shearers in the country, at times giving demonstrations at county fairs. One of them milked dairy cows, with his wife and young son handling chores when he had to be away. The Butterfield Family had one of the best Jersey herds around. The Slates family milked registered Ayrshires, raised hogs, and Mrs. Slates had a huge flock of chickens, sold eggs, and made delicious, tall, fluffy angel food cakes. The Reigle family farmed and ran a sawmill. There were older couples who were considered "retired" but who still helped their families on the farm in any way they could. So many memories of those good farm families and an enjoyable evening brought back by a plate of brownies!

With all the difficult problems that farmers are dealing with today and all the turmoil out in the world, perhaps what everybody needs is more "get-togethers", more friendly discussion, more cold milk and more brownies! We may not be able to solve the problems, but I am sure the fellowship and the brownies would help to make all of us feel better!

THE CHOICE SHOULD BE YOURS

Following a shopping trip to our local small town, the decision was made to stop at a favorite restaurant for supper. As we were seated, I noticed a gentleman that I know well seated at the table next to us. The waitress brought our menus and then proceeded to take his order, which I could hear. Liver and onions. My immediate thought was "ugh—not for me! My parents enjoyed liver and onions, however I could just never get it down, even though I tried. When the waitress came for my order it was steak, always my favorite. Even though liver is considered to be one of the most nutritious meats available, the gentleman beside me did not tell me that I should be eating liver and explain all the reasons why and I did not try to change his mind in favor of steak. We respected each other's likes and dislikes and just enjoyed our meals.

My best friend and I both love dogs and throughout our lives we have owned several different breeds. She favors Beagles and has a loveable, smart and happy beagle. I, on the other hand, prefer Border Collies and over the years have had several very smart, very loyal, and great working dogs. While we have both love our dogs as though they were our children, we do not try to tell each other which breed is the best. We respect each other's choice.

There are many dairymen and dairywomen milking, caring for and breeding dairy cattle. There are several different dairy breeds available to choose from. Fortunately, so far, we have the right and privilege to choose the breed we want to work with. I also have a dairy person in my area who chooses to milk, show, raise and sell goats. She has won numerous show awards and probably makes more profit with her goats than dairy people make right now. While I would not choose to work with goats, I respect her right to milk and raise goats just as I respect the choice of dairy breed any person chooses to milk and raise.

I have recently been reading and following numerous comments and writings about the argument of whether dairy people should be breeding a show type herd or a commercial type herd. We should all be very grateful for the right to choose not only dairy farming as a profession, but also being allowed to carry on that profession in the manner we choose. Why do some dairymen feel the need to tell other dairymen that they are wrong in the way they manage their herds? If each is happy in the way they are doing things then so be it! Respect the dairy breeder who just won a ribbon at a county fair, a breed show or World Dairy Expo. Respect the dairy breeder who is averaging ninety (90) pounds a day in his tank and has a herd making top records. Each one does the work, pays the bills, and feels the pain when he loses his best cow.

I realize a lot of the controversy comes from how to classify cows. Where is the law that says you have to classify cows? It is a choice. If you don't like the results then don't classify! If you are a "true breeder" you know which ones are your "excellent" cows. A few years ago one of the best cows in this area was purchased as a dry cow from the sale ring at a local community sale. She was registered but not classified and had no records or pedigree information. She went on to win several awards in the show ring, put milk in the tank and make some nice records. Cows that are outstanding for various reasons can be found in any herd.

Every individual has their own likes and dislikes about everything. Why is it necessary to debate what kind of herd any dairy person should breed or how they should handle them? Just respect that person for what they accomplish. Thank goodness we have the right to make the decisions about our cows! Instead of debating which way to breed a herd, all dairymen and dairywomen should be working together to find a way to keep our dairy people in business, let them be profitable and enjoy the life they have chosen on the dairy farm. Seems to me that is the more important issue at this time.

BUT ONLY GOD CAN MAKE A TREE

I think that I will never see
A poem lovely as a tree.
A tree whose hungry mouth is prest
Against the earth's sweet flowing breast;
A tree that looks at God all day,
 And lifts her leafy arms to pray;
A tree that may in summer wear
A nest of robins in her hair;
Upon whose bosom snow has lain;
Who intimately lives with rain.
Poems are made by fools like me,
But only God can make a tree.

The preceding poem is titled "Trees" and was written by Joyce Kilmer. It was brought to mind recently because of the huge maple tree that stood in the front yard at my parent's house. It was estimated to be around 140 years old. It would have been a small tree when my great-grandfather Gabriel Wagner farmed and ran a blacksmith shop in 1873. My grandfather, William Wagner was just a little boy. Perhaps it was just large enough to provide some shade as  friends and neighbors waited their turn to get their horses shod or some tools made.

The county road was built past it much later, taking out the blacksmith shop. Can you imagine the excitement when the first car came slowly chugging up the new road and drove past the maple tree, as everyone watched! My dad would have played under the shade of that tree when he was a little boy, probably with sticks and stones and bits of wood, as only rich kids had real toys. I, too, loved to play in the shade of that big tree, often with my cardboard farm and toy animals, with my faithful dogs, Stubby and Rex laying near by. When my cousin would come to visit during summer vacation we would set up housekeeping there with our dolls. It was the perfect place to quietly play, or just sit and dream about growing up. My children played there also, and have fond memories of just sitting under the big maple tree in the summer with their Grandpa and "Biddy Buddy", their nickname for my mother.

One of the things I remember most about that tree was threshing day, when the men would gather at meal time to wash-up in the big washtub that was set out, and then lay down to rest in the shade, as they awaited their turn at the dinner table. Many a Sunday afternoon was spent just sitting quietly there, resting from a week of hard work, making hay, bindering grain, plowing the weeds out of the corn, hoeing the garden, and more. Occasionally a neighbor stopped by or relatives came to visit and pass the time of day in the shade.

Every year one of the first signs that spring was on the way was the tree starting to come into leaf, and that lifted your spirits. There were always robins' nests and it was the favorite hangout for many other birds. Often in the fall a squirrel would scurry up and down before darting across the road to grab a walnut to store away for his winter food supply.

As the tree began to age, it started to split, so a huge chain was placed around it and tightened down to keep it together. It has been there for years. A few weeks ago a storm broke off one of the limbs and dropped it onto the county road. There was a danger that more of it might come down in the next storm, so a professional trimmer was called. Before he could get there, a rain storm brought with it a micro-burst that took another limb down and caused some damage. So our beautiful old maple tree has been cut down and turned into a huge stack of wood. If only that tree could have talked, the stories it could have told! So much history and so many changes had taken place around it over the years. It wasn't just a big old maple tree, it was a part of our family's lives. We are going to miss it!

HAVE YOU PRACTICED SHINRIN-YOKU LATELY?

As the hot summer sun beats down and the temperature hovers in the 90's, one of the best places to seek out is the woods. As you wander about you enjoy the cool shade of the many different trees; oak, walnut, maple, locust, wild cherry, and more. As you pause beneath a walnut tree you see lots of empty shells, and you know how that fox squirrel you just caught a glimpse of survived the winter. There goes a chipmunk rushing about. The woods are quiet except for the many different songs being sung by all the birds and the hum of the bees. There are wild flowers, ferns, and here and there some johnnie-jump-ups. If you sit down in a spot where you are hidden a bit, and stay very, very quiet, a deer just might happen along. The woods are cool, tranquil, and beautiful. Did you know when you walk in the woods you are participating in a practice called "shinrin-yoku"?

Shinrin-yoku translates as "forest bathing" or luxuriating in the woods. They are guided walks through the forest. You focus on what you are seeing, hearing, and smelling. Developed in Japan in the 1980's it requires participants to deliberately engage with nature using all five senses. The walks are often done in silence----no cell phones! Shinrin-yoku classes are taught and encourages walkers to practice deep breathing and to tune in what sparks their senses, such as the scent of wild flowers or pine cones or the texture of the bark on the trees.

By combining mindfulness and spending time in nature – two activities that have restorative properties on their own -- shinrin-yoku can yield significant health advantages. A study conducted in Japan across 24 forests found that when people strolled in a wooded area, their levels of the stress hormone cortisol plummeted almost 16 per cent more than when they walked in an urban environment. Blood pressure showed improvement after about 15 minutes of the practice. One of the biggest benefits may come from breathing in chemicals called phytoncides, emitted by trees and plants. Women who spent two to four hours in a forest on two consecutive days saw a nearly 40 per cent surge in the activity of cancer- fighting white blood cells according to one study. "Phytoncide exposure reduces stress hormones indirectly increasing the immune system's ability to kill tumor cells", says Tokyo-based researcher Qing Li, MD, PhD, who studied shinrin-yoku.

Forest bathing has become a common practice in Japan, but is only beginning to catch on in the United States. There was a shinrin-yoku group founded in Raleigh, North Carolina in 2012. People need a break from what consumes them. Nature trails are a good place to do that because you also have to watch your step, which keeps you in the moment. It is suggested by studies that just looking at green space----say trees outside an office window---- can help reduce muscle tension and blood pressure. Take a quiet stroll in the woods and enjoy the beauty and healthful benefits created there. Perhaps the answer to some of life's problems lies with Mother Nature.

BEST LAID PLANS OF MICE AND MEN
OFTEN GO AWRY

Those words written by Robert Burns applied to me this past week. My plans to go with my granddaughter to get some groceries, run a few other errands and enjoy a nice lunch together definitely were changed. After talking to her on the phone and arranging the time, I made my bed (Mom is always watching), did a few chores, took a shower and dressed for the trip to town. Suddenly it hit me, weakness, dizziness, nausea and tightness in my chest. Something was definitely wrong. My granddaughter (she is a paramedic) came in, took one look at me, ask a few questions and called the ambulance.

We are about an hour from the hospital, so it was a long ride but not as bumpy as some I have had. The EMTs were efficient and took good care of me, inserting an IV, hooking up monitors and giving me four baby aspirins. I passed on the nitro. Upon arrival at the Emergency Room I was whisked into a room and things started getting done to me. Of course, asking a lot of questions was one of the first things, even though the answers should have been in their computer. (However, once in a while there are mistakes.) I was hooked up to a machine that beeped every few minutes, so I knew I wouldn't be taking any naps! My chest was x-rayed (the board behind my back was ice cold), they took a few "pints" of blood, and I was then given a container and told to walk to the restroom and get them a "sample". (All the way I kept wanting to sing "Red Solo cup, I fill you up, let's have a party!") Upon my return to my bed, the activity ceased and it was wait---and wait----and wait.

Finally a very nice young doctor came and sat down to chat with me. After some questions and discussion, he announced that nothing serious had shown up in the tests ---- however, he felt that I should stay overnight and have a stress test. There was a look from my granddaughter, a nod from my oldest son, and since I am usually reasonably sensible about things, I agreed. Arrangements were made for me to move to the Cardiac Care Unit, which is not an admission, just an overnight stay.

The first thing they did was make me get on the scales and get weighed (having always been "chubby" I hate that!) however that brought a moment of joy as I hadn't gained back the twelve pounds I had lost last spring! Then it was into bed, hooked up to a monitor and more questions. Since I hadn't had anything to eat except a piece of toast and a glass of milk at breakfast and it was now evening, I asked if I could have something. I knew what I would get ----a cold turkey sandwich, a small bag of baked chips, a small container of fruit (mostly pears) and a cookie. (The menu in that section hasn't changed in years)! The list of drinks they offered was soft drinks and juice. When I asked about milk they said they wouldn't have any unless some was left over from breakfast. The nurse left and eventually came back with a small cardboard container of milk and for once it was cold (it usually isn't).

Naturally I couldn't sleep, so at three o'clock in the morning I was sitting on the side of my bed writing! (I always try to be sure that I have a pad and pen wherever I go). My trip for the stress test came early in the morning. All went well with the test and as soon as it was over, I was given a well- known soft drink, as the caffeine in it helps to get things back to normal following the test. My thoughts--- too bad they can't use milk. Millions of these stress tests are given every day! The two male nurses that took pictures of my heart learned a lot about dairy farming and cows, as all those hospital machines make me a little anxious, so I tend to enlighten people during those long minutes. Once in a while I run across someone who has lived on a farm but not very often. It is amazing how little most of those people know about farming and dairy products! We need to keep telling out story! After more waiting the doctor came in and reported that everything looked good and I could go home. I was grateful for the good news. While we still had no answer for what caused the "episode" I had, once I see my regular doctor I am sure there will be more tests.

One of my experiences a few years ago was being loaded on the ambulance cot directly in front of my cows in their tie stalls. They weren't scared at all, just curious and trying to lick the cot. I was in my usual barn clothes, reeking of that smell of good silage, and the other "normal" dairy barn odors. I did manage to get my "decorated" barn shoes off. The EMTs were very polite in asking if they could cut the sleeve off of my worn-out flannel shirt so they could insert the IV needle. That was my first trip in an ambulance. These days I am more experienced!

At a recent Holstein show I announced, a friend stood directly in front of me holding a beautiful three year old cow as they were being judged. A few minutes later he was on the way to the hospital with a heart attack. He was fortunate that help was close by and he is recovering. Farmers have a tendency to take care of everyone and everything else before taking care of themselves. I highly recommend that if something doesn't feel right or you are experiencing unusual problems --- don't hesitate---get help! Better to be safe than to have family and friends grieving for you!

MY BLANKET IS BLACK AND WHITE

Linus van Pelt is a character in the Peanuts comic strip by Charles M. Schulz who relies heavily on his blue security blanket. At this time there are a lot of us who are in need of a security blanket as we go through these dark, dreary, cold, snowy days and have to deal with sickness, breakdowns, low farm prices, difficult rules and regulations, taxes and more. We all are in need of a security blanket or comfort object to provide psychological comfort in our unusual and unique situations.

Security blankets can come in many different forms. At times it really can be a blanket that gives us comfort. When you finally get to the house after a grueling day nothing feels much better than settling on the couch or in your favorite recliner and covering up with a nice warm blanket. If you have a television you can turn it on to any program, doesn't matter if it is interesting or not, because in just a few minutes you will be toasty warm, relaxed and sound asleep, forgetting all the troubles of your day.

Food can be a security blanket and bring you comfort. Sitting down with the family to a delicious hot meal you often begin to relax, enjoy the food and as you talk with the family you will find yourself relaxing and the anxiety ebbing. Getting together with a group of friends at a local restaurant for breakfast or lunch and sharing problems can ease stress and provide comfort. Or call up a best friend that you can share your problems with and get together over coffee. We all need someone we can talk to.

For some listening to music or playing an instrument relieves stress. Watching a favorite sport can get your mind into the game and away from troubles. Sleeping can sometimes be a problem as the mind and brain just don't want to shut down. Leaving the television or radio on, or playing tapes can help. Some people can relax by reading. A good friend of mine always ate a big bowl of ice cream every night before going to bed. He loved ice cream and the pleasure of eating it relaxed him.

Many people rely on their strong faith to get them through their troubles. My grandmother was one who always relied on God to see her through her troubles. She had a lot of them to deal with including a health problem from childhood that there was very little help for and no cure. She never stopped believing no matter the challenge she was given and the strength she had was unbelievable. She never gave up!

Young children need a teddy bear, a doll, a stuffed animal, favorite toy or a special blanket to give them comfort and relax them so they can go to sleep. Due to the many things adults deal with every day of their life, they also need some type of security blanket to comfort and relax them. However, they often don't want to admit it. According to some research a significant per cent of adults still sleep with teddy bears. There are many items that represent emotional attachments and it is very common for people to carry items with them and look to them for emotional support. There are times when we all need a security blanket and comforting. Don't be afraid to reach out for it when you are weighed down by troubles, trials and tribulations. It is not a sign of weakness. Find that security blanket that helps you get through each day and like Linus hang on to it! Any type or color will do. Things will get better!

LOCAL SALES HAVE SOMETHING FOR EVERYONE

The coming of spring results in a beehive of activities known as garage, yard, or barn sales. Spring cleaning usually means finding things that are no longer needed, clothes that have been outgrown or just the desire to get rid of some things. These sales are so important for young families with growing children in this day and age. Raising a family costs so much and children outgrow their clothing so quickly. Prices for new clothes and shoes are so high! There are so many different things at these sales, you just never know what useful item you might find. Usually everyone can find something of interest that they just can't pass up!

The first yard and garage sales actually got their start in the shipyards in the 1800"s. They were called "rummage sales" where the shipyards would sell unclaimed cargo at discounted rates. They became popular in communities all over the country during the 1950"s and 1960's. In my early years of marriage, yard sales had not begun in this area. We would attend household auctions and purchase boxes holding various kitchen or household items for a dollar or two. We didn't have much money to spend so the auctions were a great help in setting up housekeeping. I still have a few things I use from those years!

On a recent afternoon of "adventures" with my granddaughter, we stopped at our local county home where a barn sale was being held. There were two buildings filled with items to sell, all donated by people in the community, and with all the proceeds going to help maintain our county home. There was such great support for the sale and for our county home! These sales are a great way for many organizations to acquire funding.

For the ordinary family the proceeds from a yard or garage sale can mean extra money for a vacation, needed household items and much more. These types of sales are a benefit to both seller and buyer!

We also stopped at our local "Browse and Buy". I had some things to drop off, as I had often done before, however I had not been in the store since it had moved to the current location. The store has lots of neat, clean rooms, all filled with quality merchandise of all kinds. What an asset it is to our community! The store is run by volunteers and the proceeds go to Community Hospice Patient Care. It is so interesting to just walk around looking at all the different items that are on display. There were many beautiful things! I was pleased to find a lid to fit my favorite crockpot, as the old one had been lost while on loan to someone. And, of course, I found a couple other items to buy. I just couldn't pass them up!

These types of sales and stores are important to our community. Not only do they benefit organizations and people, they also provide a place for people of all ages to purchase items that they might not be able to find or afford elsewhere. Take the time to support your local sales and stores. Spending time there looking at everything can be fun! You often find some surprises!

MAILMAN, MAILMAN I WATCH FOR YOU EVERY DAY

"I can always depend on you whether skies are blue or gray". I always look forward to the mailman coming. It is very unusual for my mailbox to be empty. My mailman probably wonders why I get so much stuff! Before all of today's technology came about the mail was my way of getting the news. Years ago I wrote numerous letters to friends and relatives and it was a happy day when I received one in return telling me all the family news. A magazine in the mailbox such as Good Housekeeping, Reader's Digest, or the Ohio Farmer, or new catalogs from Sears and Roebuck, Penney's, or Montgomery Ward were a delight. And, of course, being involved with registered Holsteins meant receiving the Ohio Holstein News, Holstein World, and several sire directories. I spent hours reading them and researching information about the bulls I wanted to use in my herd. I love to read and I subscribe to several popular magazines, local newspapers, and all types of dairy magazines. When something interesting comes in the mail I grab a cup of tea and enjoy!

So much of the mail filling up my mailbox these days is different and they almost all want me to order something or send for more information! Among the things in my mailbox today was a huge envelope from that "certain type house" telling me that I can win a huge sum of money, especially if I order one of their products. To win extra money there is a "scratch-off card" to include in my order and there is my "lucky bingo card" to win even more. They are offering "can't miss free deals plus huge savings". Among my choice of things to order is a novelty train clock that will amuse me every hour on the hour with authentic train sounds and flashing lights. Just what I need to keep me from enjoying those quick naps I take during the day! There is the "Energizing Crystal Feng Shui Luck Charm", which will bring light and rainbows into my life, boost positive energy and decrease negativity. I can hang it in a window or carry it with me to keep my spirits energized. Maybe if I had one of those it would make me more enthused about cleaning the house on these dark and dreary winter days. I should probably order one of those instead of the "satin caftan that's perfect for lounging"! And then there is the hand crafted green jade good luck elephant. There is also a "Learn to Play Piano in Six Weeks or Less Book", but I really should have a piano if I am going to order that. After all I can order any of these things for only four payments each! And I am warned that "failure to properly enter will result in automatic forfeiture of the "Forever Prize"!

Perhaps one of those "Perfect Sleeper Easy Care Pillows" could help me get to sleep at night instead of lying awake for a couple hours and then getting up and writing down all those ideas for my column that have been keeping me awake! Oh, there's another sweepstakes advertisement. I have six chances to win an "Easy Rest Adjustable Bed". I don't have to purchase anything, just give them my phone number so they can provide me with information. Right! Maybe I should enter and try to win that bed, then I could take my "Perfect Sleeper Easy Care Pillow", hop into that bed, and maybe I could go to sleep at night instead of writing my column in the wee hours of the morning! One thing is for sure, sending an order or information will guarantee that my mailbox will become fuller in the future! (I'll let you know if I win the bed!!!)

SEND MONEY AND THE CARD
TO RECEIVE YOUR FREE GIFTS

I have always looked forward to the mailman coming and finding something interesting in the mailbox. Years ago calling long distance on the phone could cost a lot of money, so there would often be a card or letter from relatives or close friends with family news and often pictures. In the evening after chores were done, I would sit down to answer those letters and let everyone know what was happening in our family. As technology has changed our ways of communication, the personal type mail and letters have almost ceased to exist.

However, I still seem to get lots of mail, most of it asking me to send money or send in the card for the free gifts and then they will let me know how many payments I will have to make! There recently was an ad from a certain society wanting me to help save the birds. If I sent in a donation, I would receive two free tote bags with very ugly birds pictured on them (why not bluebirds, cardinals, or robins) and my very own membership card would be included.

Another day there was the offer for a "newsletter for caring cat owners". This was a publication that said "cracking the code of communication with your cat can be the key to enjoying a healthier and happier life together". There has always been cats here on the farm, they have always lived in the barn, caught mice, drank milk, were supplied with plenty of cat food and seemed to be happy. I really didn't worry about communicating with them, just petted and talked to them. I really didn't feel that understanding them was the most important thing on my list of things to do. Most of them lived to an old age, including one we referred to as "Snaggle Tooth", as she was so old her teeth had actually fallen out. Years ago when we hand milked it was always fun to squirt milk at them and watch them sit up and catch the stream of milk in their mouth.

There was an ad for the "Tribute to the American Dairy Farmer Special Edition Marlin 1895G 45-70 Caliber Rifle". Only 400 were to be made to pay tribute to the American Dairy Farmer. According to the ad, "it serves as a symbol of our history and heritage". I don't know why they thought a dairy farmer needed a gun, but I am sure there are a lot of milk inspectors who wouldn't support this advertisement!

There is an ad for me to purchase 12 charm bracelets, one properly made for each month of the year. I can also get free gifts and if I don't want to wear June's bracelet in June it is perfectly o.k. for me to wear it in December. If I can wear any month's bracelet at any time, why do I need 12 of them? No matter when I wear them I am going to look fashionable and festive! I don't even have to send any money ---- now.

I am constantly getting letters wanting to sell me a warranty on my car. According to them, I just shouldn't be without it and all I have to do is send the money. Are they really going to give me a warranty on a 1981 Ford LTD? There are the offers for insurance, all kinds, and even though I am sitting on one of the high hills in Carroll County, there is still the option for flood insurance. Believe me, if a flood gets to me here, a lot of people are going to be in big trouble!

And then there are the catalogs! I made the mistake of ordering from one and now I get a catalog just about every day! There are catalogs for me to buy items for my dogs, my birds, my fish, my cows, my house, all of the people in my family, neighbors, and complete strangers! And there are all those magazines you can send subscriptions to, including the ones that offer you a chance to win money in a big sweepstakes. The large print tells you that you don't have to send money, but the small print tells you how many monthly payments will have to be made. And of course, when the renewal is due you don't have to do anything, they will just keep sending them and bill you! The one thing all these offers have in common is wanting money. If these companies knew anything about farmers, they would know there is never room in a farmer's budget for unnecessary things. Farmers have to make difficult financial decisions every day and most have a wastebasket close to the area where they open their mail.

    A trip to the mailbox is always an adventure for me, as I never know what I am going to find. On occasion during the summer it has been a bag of delicious red ripe tomatoes! No one asked for any money!

THE RETURN OF THE MILKMAN

Years ago as I rode the school bus through the small town of Lamartine (the name was later changed to Perrysville) a familiar sight would be the milkman, Ray Umpleby, delivering milk to the local residents. The empty milk bottles would be waiting on the steps of the houses with money inside. Mr. Umpleby would leave the full ones and retrieve the empty ones to be refilled for the next delivery. He was a dairy farmer who lived on a very clean and neat farm at the edge of town. The milk came from his small herd of beautiful Jersey cows.

I recently read an article that asked, "Could the milkman be coming back?" In the 1950's more than half of the milk produced was home delivered. Much of our country had the service. With the current trend for consumers to order so many different things over the internet, could ordering dairy products by way of the internet become popular? Consumers order all types of things on the internet and have them delivered. In some areas groceries are being ordered on the computer and delivered directly to the homes and the consumer makes no trips to the supermarket. Why not make dairy products available to be ordered from the processor or special dairies and delivered to people's homes just as they used to be?

There are already a few companies who do deliver and they say their customer base is growing. Many of the suppliers are small, family owned dairies providing non-homogenized and low pasturized products, using both plastic containers and old style glass bottles. The cost of the milk is higher but the consumers are willing to pay the price for a product that tastes better. There are different ideas including memberships allowing consumers to choose which dairy their milk comes from. Deliveries are on a set schedule. Some companies offer other items including all types of dairy products, fresh produce and more. The old service of delivering milk is becoming new again!

Dairymen and dairy women are constantly being told that we need to find ways to encourage the consumers to buy and use more milk and other dairy products. Having those products delivered would save today's busy consumer the time it takes to run to the local store to pick up that needed gallon of milk or other dairy product. Perhaps more would be consumed by the family if the milk, cheese, yogurt, butter, cottage cheese, ice cream and dairy treats, arrived at the home on a regular schedule. The kids would be delighted to know when the ice cream bars, Snickers bars and other dairy treats would be arriving!

While we do have to recognize that dairy products require refrigeration, in this day and age there are many ways to take care of the products and there is no doubt in my mind that the consumer and the milkman would work together to solve that problem. Consumers are constantly letting us know that they want more local, organic, range free and grass-fed products and they are willing to pay for convenience. We need to sell more milk and dairy products! Perhaps it is time to bring back the local milkman in more cities and towns!

RESPECT MOTHER NATURE AND ALWAYS TAKE TORNADO WARNINGS SERIOUSLY

Lately it seems to rain on the days I choose to write my column and that is exactly what it is doing today. The rain is pouring down, thunder is booming, wind is blowing, lightning strikes are sharp, and we are under a tornado watch. I always take tornado warnings very seriously, having gone through a tornado several years ago.

It happened just about this time of year. The storm came up very early in the morning, just before daylight. My husband was working midnight shift, and the children and I were alone in bed. I was awakened by the thunder and by my border collie, who hated the sound of thunder. I went around the house shutting doors and windows. I could hear the sound of the storm getting worse. I raced upstairs to get the children and seek shelter in the basement. The tornado hit before I could get downstairs. I stood petrified with fear at the top of the steps, with our baby in my arms, as it came roaring down over the hill and tore through everything.

When people tell you that it sounds like a freight train, believe them, it is true. However, it is a sound that you will never forget. The roar only lasted a few minutes and then it was gone. There was only the sound of the rain. The four car garage that stood between my house and the house of my parents was completely destroyed. Yet hard to believe, neither house was damaged. Trees and electric lines were down everywhere. As it was still dark, my Dad went out with flashlight in hand to try to access the damage, and to make his way to the barn to see what was damaged there. He cautioned my mother to stay in the house and yelled for me to do the same, as we couldn't tell where all the downed electric lines were laying. The cows and heifers were all out on pasture, so we could only hope that lightning hadn't struck where they were or a tree gone down on them. There were baby calves in the barn.

As daylight came, we could hardly believe the damage. A part of the roof of the garage lay on top of our new pickup truck and my parent's car. Items from the garage were scattered up through my yard, in the barnyard, and all through the fields. Later on we found items in the neighbors hay field a half mile or so from us. There was a huge solid oak post about ten inches square that stood in the center of the garage as a support for the structure. Attached to the post was a huge and very heavy drill press. That post now lay several feet from the garage in my yard. Yet an open umbrella that was made to fit on our Oliver tractor stood in its regular spot and hadn't been moved! And there was a piece of straw embedded in a two by four! There were several things about the destruction that had no logical answers!

The electric company and neighbors soon came to help. The barn just had some minor damage, and there were some trees down. The livestock were all safe and we were able to milk the cows and cool the milk. Everyone in the family was safe and we were very thankful for that. So, when I hear that we are under a tornado watch, it brings back a lot of memories of the destruction and of that sound. If a storm gets bad, I do not hesitate to head for the basement or look for a safe place. Make sure you have a safe place to take shelter and do not hesitate to do so when there is a chance of a tornado. Never take the storm and tornado warnings lightly. Mother Nature can get pretty angry at times and we should always respect her.

JUST DOING WHAT I CAN TO
KEEP PEOPLE WORKING

It seemed like a simple request ---- all I wanted was a faster computer. When I first got a computer, the only service available down here in the hills was a dial-up service. Not only was it very slow, but it also tied up the phone line when I was on the computer and I couldn't receive phone calls. Cell phones don't work in my immediate area. So, my eldest son who takes care of all my computer problems, contacted a company that provides both television service and computer service. A date and a time was established with the company and my son made arrangements to be here to supervise the installation. The time was set for 11:00 a.m. on a Thursday.

On Wednesday morning there appeared an unknown man on my front porch. When I asked what he was doing, he informed me that he was here to install the television service. When I explained that he was here on the wrong day, he politely informed me that he would go ahead and do the installation. I then politely informed him that due to the fact that my son needed to be here to supervise where the holes were being drilled and the wires were being strung throughout this old house, he would have to wait until the appointed day. Since his cell phone wouldn't work here, he asked to borrow my house phone to call the company. After two or three calls, he established that the company had no record of ever having talked to my son, but had tried to get in touch with him on a wrong number. Even though there was no record of the appointment having been made, somehow this "television service installer" had magically appeared on my porch! He headed for his truck, grumbling something about the fact that he probably wouldn't be the one coming back.

The next day at the appointed time, another "television service installer" came to the door. My son had already arrived and showed him where things needed to go. He was a very nice man, very careful about his work, and I hoped the company appreciated what a good asset he was to their business. The "computer service installer" was supposed to also have been here at the same time, however he didn't show up until several hours later. He informed us that he planned to install the device for the computer on the roof. We informed him that we did not want it done that way, as it could cause the roof to leak. He then told us that there would have to be a pole set and that he didn't set poles. He said he would notify the company and we would have to wait for the "pole setters".

A few days later I observed a strange man walking around the house. I assumed he was a "pole setter". I was wrong. He was a "pole installation inspector". His job was to determine that there were no electric or phone lines buried in the ground where the pole hole was to be dug. Since the lines were all at the opposite end of the house from where they would be digging, there was no problem. He was a very nice young man, and we spent a couple hours talking about how nice it was out here in the country, deer hunting, and a few of the world problems. He then left to go to lunch.

A few days later another "computer service installer" knocked at the door and asked where the pole was located. When I told him that no hole had been dug and no pole had been set, he told me that someone would be back later, but probably not him, and he went on his merry way!

A week or so later, the "pole setters", a man and a woman, arrived with a pickup truck pulling a trailer, upon which there was a back-hoe. They had all sorts of equipment to dig about a six inch hole for a three inch pipe. The only thing they had to borrow from me was a bucket of water to mix the cement. Once they finished, all I had to do was wait a few more days to let the cement set up.

Finally another "computer service installer" arrived and went to work to change everything to a fast computer. After I explained that there were certain holes the wire was supposed to go through and I didn't want him just stringing it everywhere, as he started to do, he went about his work and after a few hours I finally had faster computer service. It only took almost three weeks and eight (8) different people to accomplish that. I am just thankful that I can do my part to provide work for so many people and help to keep the economy improving! Now all I have to do is learn how the new faster service works.

PLAY FOR THE LOVE OF THE GAME

I am not a baseball fan! So what was I doing sitting up until 2:00 o'clock in the morning watching game five of the World Series as the Houston Astros finally defeated the Los Angeles Dodgers 13 to 12 in ten innings? I am a football fan and I had been watching one of my favorite teams win their game. After it was over, I was flipping through the channels and I happened onto the World Series just as some player I didn't know hit a home run with a couple of his teammates on base. The crowd was roaring, the announcers were excited, someone kept blowing a train whistle, and everyone was waving orange towels. I guess I just got caught up in the excitement! Both teams kept getting runners on base and then someone would hit a home run! The lead kept going back and forth until Houston finally won. Then no one wanted to go home! They just kept hugging each other, jumping up and down and waving their towels! They had played baseball for 5 hours and 17 minutes! I went to bed.

Then there I was again --- watching game six --- I was hooked. There were some interesting network interviews with some of the young players that I had never heard of before. The game began, one of the Astros hit a home run (I didn't know him either) and they took the lead. For a while not much seemed to be happening, so I took a little nap. When I woke up the Dodgers had taken the lead and they would go on to win a much quieter game than game five. There wasn't so much excitement and I had time to pay attention to the players. They do some "different" things! They seem to scratch different places, they fuss with their hats, and pitchers seem to take a lot of deep breaths. The players waiting in the dugout are interesting.

Some of them were chewing bubble gum and blowing bubbles. One player took a huge wad of gum from his mouth, looked at it, and then put it back in! Some of them spit out some type of seeds. One fellow seemed to be waxing his moustache. The haircuts are interesting --- some of them looked like someone had partially clipped them and forgot to set the topline! The game finally ended and the announcers kept stressing the importance of game seven --- that it was the only game that counted. If that was the case, why didn't they just play game seven and forget about the other six games!!

Yes – I watched game seven. It was a good game and the players gave it their all, both teams should be very proud of their accomplishments. The Houston Astros won their first World Series. They would take home the trophy and everyone was so excited. It was interesting to see the tears of joy in the eyes of those big, tough baseball players. Babies were being tossed in the air and there was a marriage proposal. She said "yes"! Game seven brought out many emotions.

Years ago we had our own baseball diamond just down the road from our farm. Every summer after church and dinner neighbors would gather there to play baseball. The players were of all ages, from the youngest who was just learning how to play to the oldest who was still able to play. Sides were chosen. Someone was designated to be the umpire, and the game was on. My Dad loved to play baseball and he was a left-handed first baseman. I was always told that playing first base was harder for a left-handed person than for a right-handed person. My Dad was a pretty good player and hitter and he really enjoyed those Sunday afternoons. Our neighbors had eight children --- seven boys and one girl--- and they were really good players, as they practiced at home. The daughter was just as good as the boys and a little faster running the bases, so she was always chosen quickly! Everyone who didn't play brought their lawn chairs or sat on make-shift bleachers --- boards laid on cement blocks. Friends and neighbors visited and shared their news as they watched the game. The younger children played in the dirt with their trucks and farm toys

George Wagner in his baseball uniform - 1928

or played games. Some players were really good at baseball --- some were terrible --- but everyone had a lot of fun! During the World Series the many troubles of the world were forgotten, as people came together to enjoy a baseball game, just as they did on those Sunday afternoons long ago. In baseball as in life, all the important things happen at home.

HORSIN' AROUND ON A RAINY DAY IN MAY

The first time my Dad held me up to pat old Ted's nose, I was hooked on horses. Ted was a big dappled gray draft horse, strong, slow and gentle. One of the things I remember most about him was my Dad talking about the size of the horseshoe he needed, a size eight, as he worked at putting new shoes on him.

Ted was getting older, so Dad decided to buy another horse to work with Ted's stable mate, Dan. That was Belle, a beautiful Belgian mare with a big white blaze down her face and four white socks. However my Dad was "scammed" by the seller, as he didn't disclose that Belle had "fistula", an infection in her withers that affected her ability to be worked. Dad decided to breed her and raise a foal with her, so along came "Jerry". That was the one and only time that my Dad gave me a "whippin". I was so excited about the baby colt that I

Dan & Prince

wouldn't stay out of the box stall. Dad finally had to use a peach tree switch to get me to listen. Belle wasn't mean but Dad was afraid she would accidentally step on me----I was only four years old.

As soon as I was big enough for Mom to allow me to walk

to the fields where Dad was working, I would walk a mile just to get to ride one of the horses back to the barn. As soon as I was old enough, I started driving and working the horses. It was a happy day for me when Dad would let me stay home from school in the spring to work down the plowed ground in preparation for planting.

The next horse that came to our farm was Prince, another Belgian. I went along on the day Dad bought him from a local farmer. Prince was all "business", eager to work and a little more anxious to get going than his partner, Dan. Dan, a big bay, never got excited, but was tough and could work all day without seeming to get tired.

On Sundays I spent hours riding everywhere on Prince. Our horses were big and well fed, so it was like riding a barrel and, since I usually rode for hours, sometimes when I got off I could hardly walk! I, of course, rode bareback and only got thrown off once----when Prince accidently switched his tail into the electric fence. I saw a few stars but never let my Dad know as I was afraid he wouldn't let me ride him anymore! I always wished for a riding horse, a Palomino like Trigger, however it was never in the budget.

Dad bought his first Oliver tractor so he could get the work done faster and for a while we worked both tractor and horses. However the time came when Dad realized it was time for the horses to

go in favor of another tractor. It was a hard decision for him as he loved his horses, but he knew it had to be done. I was heartbroken. On the day the dealer came to take them I put baby daughter Cindy in the baby buggy and walked to the far side of the farm, sat down in a field and bawled. I still get tears in my eyes when I think about that day. My head knew it was the right decision but my heart didn't want to accept it. After that day I traded in my love of horses for the love of my Holstein cows.

My granddaughter, Kristin, keeps three horses here, Jessica, her mare she has had for several years, Skip, who looks like a buckskin but doesn't have the black mane and tail and Rayna, a sorrel quarter horse with a white blaze in her face. I can't ride them but I enjoy watching them in the pasture. Perhaps someday great granddaughter Emily will ride that Palomino.

I often watch racing on television and enjoy it even though I don't know anything about the horses that are running. Like so many others, I enjoy watching the Kentucky Derby and everything that leads up to the actual race. Seeing the horses close up and hearing about them is so interesting, including the prices paid for them. And there are interesting stories about the trainers and owners. The hats and the flowers are great! I was watching all that last Saturday and enjoying it so much ----- until the end! It was pouring rain, a horrible sloppy track, too many horses, all of that creating a rough race and things were bound to happen. In my opinion (which doesn't count) the fastest horse won. There will be arguments and turmoil about this race for years to come. How sad for horse racing's special day, for horse lovers, for everyone involved, for the owners and for a courageous and beautiful horse. The 2019 Kentucky Derby will never be forgotten!

CAN WE GET THAT CHARLEY HORSE BACK ON THE TRACK?

Sometimes you will find me just horsing around! One of the things I enjoy doing in my leisure time is watching horse races. The Kentucky Derby is, of course, the favorite. Seeing all those beautiful thoroughbred horses, hearing the information about them, learning their pedigrees, plus all the pageantry, the people, the hats and the excitement of the race is great. After that comes the Preakness and the Belmont Stakes and then the Breeder's Cup. There are races on a television channel that I can turn on and watch anytime. They take place at tracks all over the country. There is very little time between races and if there is a commercial it is usually for a magnificent stud horse.

I enjoy all the different and odd names of horses. I have learned that naming a thoroughbred horse is complicated. There is a long list of rules and regulations that must be followed. There are certain names of horses that you are not allowed to repeat and all names must be approved. I once named a new baby calf "Lucky Charm" after the horse that had just won the Kentucky Derby, however I didn't ask anyone's permission. My veterinarian, who owned race horses, saved the programs from the races so I could check out the names and see if there were any I wanted to use for my Holsteins.

The most important thing with any type of horse, be it a race horse, draft horse, quarter horse, buggy horse, or any other kind, is keeping it healthy and able to perform its task, whatever that might be. I recently learned of a device being used to help with lameness, injuries and whatever is needed to keep race horses healthy. It is called an Equine Vibration Platform and is being used for whole-body vibration to help cure lameness. The horse is simply lead onto a large platform that literally vibrates. It is said to help with a wide variety of health issues, from increasing blood flow and range of motion to improving joint health and reducing inflammation and pain. Research is currently being done to determine the effect of the treatment. As with treatments for any health problem, there can be other factors that enter in to the success or failure.

It is a well- known fact that we senior citizens quite often deal with some type of "hitch in our git-along" and sometimes we get a "charley horse"! We have to find and try different ways to help with our problems. Just this morning I was dealing with some pain in my "hock", evidently I twisted it a little bit yesterday. There is a multitude of us with back problems --- especially those of us involved in farming. Problems caused by the bales we tossed around, sacks of feed lifted, buckets of milk carried, manure pitched, holes dug, and much, much more. There are the ailments from being run over, tossed off of, stomped on, squeezed, butted and jerked around by various kinds of animals. Fall is here and our bodies are beginning to feel the changes in the weather.

If the whole body vibration treatment proves to be of help to race horses, perhaps it will prove to be of help to some of us! If we have trouble standing up on the platform, we can always use our canes and walkers, and if it proves successful maybe we can throw them away! Who knows what research can accomplish in the future for both animals and humans! We have come so far in the last few years!

A TRIBUTE TO THE LATE PAULY OPOSSUM

Living out in the country means you will see lots of wildlife. In my area there are deer, coyotes, wild turkeys, groundhogs, fox, pheasants, rats, raccoon, and lots of birds of many kinds including the hummingbirds that arrive in the spring and leave in the fall. So far I haven't seen a bear, thank goodness! The wildlife seldom cause any problems for me although an occasional groundhog will move in making it necessary for me to treat them to some "double bubble".

A couple months ago, during the still dark hours of the morning, I was unable to sleep, so I went out to sit on my deck. I heard a little noise in my yard, so I picked up my flashlight and looked around. There he was---furry, long tail, hairless ears, dirty white in color, just munching away and paying no attention to the light shining on him, an opossum!

The kitchen in my very old farm house is not very modern and I do not have a garbage disposal nor do I own a pig. I do not want to put peelings, leftovers, etc. in my wastebasket, so I have a place in the backyard where I dump them. "Pauly Opossum" has been coming to a late supper every evening and enjoying whatever is on the menu. He isn't hard to please and seems to especially enjoy muskmelon rinds and the left over salt and butter from the sweet corn cobs. He has a healthy appetite and seldom rejects anything on his private buffet.

I really don't care much for opossums, as they are not the cutest and look so much like a rat. Rats will send me screaming! However, he wasn't really bothering anything and since I don't have any pets, he was welcome to clean up the items I toss out and I decided to refer to him as "Pauly". I assumed he (maybe she) had a family nearby, however I never saw anyone one but him (or her).

I decided to tell people I had a new "pet", however I am not sure I could have taught him any tricks. There used to be one that acted in an ad on television but all it did was play dead! I wished he was a more attractive animal, however he was working hard at keeping my yard clean. I was never considered a "looker" either, but I like to think that people were happy with me, as I worked hard and did my work well. My grandma always said, "Beauty is as beauty does. Looks are only skin deep". Grandma had a lot of sensible sayings!

The scientific name for opossums is "Didelphis Virginana" and they are located mainly in the Eastern United States. The "O" is silent, so we generally call them "possums". Males are called "jacks", females "jills" and the young ones "joey". A group of them is called a "passel". They are smart and have fifty teeth. Opossums really are very important. They eat rodents, insects, snails, slugs, birds, eggs, frogs, plants, fruits, grains, table scraps, dog food, cat food, cockroaches, beetles and more. They are allies in fighting Lyme disease. A single opossum may kill an astonishing 5,000 ticks in a week. They clean up ticks, venomous snakes, discarded bird seed and more. They are sanitation workers of the wild. Of all the wildlife visitors to your back yard opossums are one of the best to have around.

Alas, when I arose a few mornings ago and looked out my sliding glass door, Pauly lay in the middle of the road with eight buzzards surrounding him and delivering the eulogy. I felt sad and I will miss his visits. The next time you see a possum in the road, please remember Pauly. Try to avoid it and don't run over it, even if it is just lying there --- it could be practicing acting! They are an important part of our environment.

A TALE ABOUT A FARM DOG

When I was a little girl growing up most farms had a dog. On some farms it was a dog that was specially bred to be a "stock dog", whose job it was to herd various types of livestock. On other farms it was just a "farm dog" to herd livestock, catch groundhogs, play with the kids, be a "watch dog" when someone came around and more. Our dog was a bobtailed shepherd named Stubby. My dad purchased him from a local breeder. I never quite understood why the pups were bobtailed, but he had a short tail that he could wag.

The cows were out on pasture and had a large area to roam over. There were times when they didn't come home at milking time. In those days they didn't have all the feed, silage, and TMR waiting for them that they receive today. Dad would take Stubby and go to find them and I would tag along. The pasture had some woods and grazing areas that were pretty far away. Once Dad spotted the cows, he would make sure Stubby saw them and give him the command to "go get the cows".

Away Stubby would go, down the hill, across the creek, up the hill and through the field where the cows were eating. He would run right up to the cows, look all around, then turn and run right back to where we were standing. Dad would take hold of his collar, shake him a bit, and in a more stern voice command him to "go get the cows". Stubby would then bite my Dad, not a hard bite but a nip, and take off as hard as he could run. Once again, down the hill, across the creek, up the hill, through the field right up to the cows. Then he would bark and start the cows for home. Usually two trips would do it but once in a while it took three!

I had been allowed to adopt a puppy from a litter that my

aunt and uncle had. He was a Cocker Spaniel and Coonhound cross. He grew up to be the size of a hound and jet black. I named him Rex. Stubby and Rex usually went everywhere with me. I loved the team of draft horses that my Dad drove. I would follow everywhere they were being worked and dream of the day when I could drive them. I decided I could have my own "team". I gathered up some rope clothesline and fashioned my own harness to fit Rex and Stubby. I drove my "team" everywhere. I had a small wagon that I hooked up to them and then I could haul "stuff". That worked well until one day when a neighbor came up the road with his horses hitched to a grain drill. When the dogs saw him, away they went, down through the yard, over a stone wall, barking and dragging a rattling wagon all the way. Fortunately the neighbor had a very docile team of horses, who while startled, didn't run away! I was crying and so mad at my dogs for not listening to me and the neighbor was doubled over

STUBBY & REX

in laughter!

My Mom loved chickens and she always had a big flock that ran free everywhere. Of course, there was a big old red rooster and he was mean. He would run at you, attack and "flop" you every chance he got and I was afraid of him. One day I went out in the yard to play and he took after me. I started to run and scream and out of nowhere came Stubby. He ran right over that old rooster, both of them tumbling over and over, with grass and hair and feathers flying everywhere. When the dust cleared, the rooster was running for the chicken house and Stubby was standing looking at me, spitting red feathers out of his mouth and wagging his stub tail. He was no longer just an ordinary farm dog, Stubby was my hero!

ROCKIN' WITH THE HUMMINGBIRDS
ON A BEAUTIFUL MORNING

The hummingbirds are back! The first one showed up last week. I found the feeders from last year, filled one of them up and hung it out, only to discover it was leaking. Since we had two, I filled the second one, only to find that it was also leaking. (So much for plastic stuff made in China!) So it was off to our local Tractor Supply to purchase a new feeder. They had several to choose from. After much indecision on which one to get, I settled on one that looks like an old fashioned canning jar. It has a square wire handle to hang it by, similar to the jars from years ago that had "wire bails". I came home, made the nectar, filled the feeder and hung it on the front porch.

The next morning was a beautiful and warm morning (no rain!), so I took my cup of coffee and went out on the porch, sat in my new rocking chair that my son gave me, and watched the hummingbirds feed. It was so peaceful and enjoyable. The dogwoods and lilac bushes are so pretty. All the colors of spring are so enjoyable after a dull, dark winter.

The following morning there he was -- Bullybird! He was sitting on the handle of the feeder surveying his kingdom! Every time another hummingbird came in to feed, he went after it and chased it away. He reminded me of those old war movies where our airplanes would just be flying along in the sky and suddenly an enemy airplane would dive down and attack, and our planes would then scatter in all directions. Bullybird is a "dive bomber"!

So how do you stop a Bullybird from chasing all the other hummingbirds away? When I was a kid in school and a bully picked on me I just "socked" him. (You didn't get suspended from school for defending yourself back then.) Obviously the other mockingbirds were too afraid of Bullybird to engage him in a fight. The answer had to be another feeder in a different location.

On Saturday, a call from my oldest son telling me that he was coming down to the farm and would be taking me out for the day, meant that I would be planning on stopping somewhere to buy a feeder. However, I didn't know that our stop would be the Hartville Hardware. What a place! If you don't find what you want there, you probably won't find it anywhere! An entire day could be spent just looking at everything. My one and only purchase was a hummingbird feeder, however there were a couple of other items that definitely interested me, including a set of dishes that looked like Holsteins! My advice, don't go to Hartville on Saturday! The amount of traffic and people is unreal! People travel for miles and miles to go to the Hartville Flea Market and to eat at the Hartville Kitchen. There are many other attractions. Lunch at Grinders was very good after a wait to get seated. Then it was on to my son's house to visit with family.

Our drive to Hartville was on some of the less traveled country roads, the type of drive that I always enjoy. Stark county fields are very wet and few have been planted. A lot of them have standing water. Spring is proving to be as difficult as winter for farmers. I wish more farms would put up signs telling who they are, as I saw several with Holsteins but had no idea who owned them.

This morning I hung the second feeder out and was rewarded by the birds coming to feed immediately. In just a few minutes there he was, perched in his spot on the first feeder. Meanwhile the other hummingbirds were enjoying themselves at the new feeder around the corner. Bullybird is going to be kept very busy if he tries to boss both feeders! Who knows, more birds are showing up and one of these days he may just meet his match!

THE WILDLIFE ARE KEEPING MY LIFE BUSY

One of the pleasures of living in the country is being able to observe and enjoy wildlife. Lately it seems the wildlife are taking over my life! I always look forward to the return of the hummingbirds in the spring and I enjoy sitting on my deck with my morning cup of coffee and watching them feed. These days they are keeping me busy making nectar and keeping the feeders full. They are consuming a half a gallon every day! I have two feeders that I put out, one is larger than the second one. The smallest feeder has to be filled twice a day. I have counted ten hummingbirds at the feeders at one time and I have no idea how many others were waiting their turn! I have one little bird who comes right up to the sliding glass door and hovers there. When I see him doing that I know that the small feeder is empty and he is looking for a refill! Thank goodness sugar has been on sale recently!

One morning I walked out on the deck with my coffee and looked down into the front yard. There was a huge buzzard marching back and forth like a commanding general while three more sat on fence posts across the road just observing. A car had hit an opossum and it was laying in the yard. So I had to get a shovel, find a burial spot, dig a hole, and say a few words. I like decorations in my yard but buzzards are not one of my choices!

The next bit of wildlife I discovered was ants! One morning I found them crawling all over the sink in my kitchen. Usually if I have an ant problem I find them coming in around a window. We never did find out where these were coming in, as the sink is located in the middle of the kitchen. So it was a trip to town for ant traps and very careful handling of food and dishes until I got rid of them.

Then it was a raccoon! Even though there was nothing for him to eat, he decided that my deck was a good place to play in the early morning hours. Of course he left his "calling card" after every adventure! So there was another trip to Tractor Supply for an animal trap. The first night my granddaughter fixed a slice of bread covered with peanut butter, placed it in the trap, and my grandson-to-be set the trap. The next morning I discovered Mr. Raccoon had enjoyed the bread and peanut butter but the trap was empty! Something hadn't worked right and the trap didn't close. However, after some adjusting and more bait, a few mornings later there he was waiting for his vacation trip to another area!

Having a yen for pasta, I went to the cupboard to get a box of ziti. When I opened it I noticed some of it didn't look right and after further investigation found little black bugs in the box! Lo and behold, all my pasta and flour contained these little black bugs! So everything had to come out of the cupboard, several food items had to be thrown away, and the cupboard got a thorough scrubbing and cleaning! I am hoping that took care of them.

The current wildlife problem ----carpenter bees or wood bees in our pole buildings. The bees bore into the wood and deposit their larvae. The woodpeckers have come in, made huge holes in the rafters to get at the larvae and in doing so have weakened the supports in the buildings! The results----- we will have to either repair or replace the buildings! I can't help but wonder what my next encounter with wildlife will be, but rest assured that when driving into town I will be on the lookout for deer!

SPENDING THIS WEEK WITH LILY

It was the kind of phone call that most grandparents receive at one time or another. My youngest granddaughter was on the other end of the line. After exchanging a few pleasantries came the question, "Grandma, can you babysit for me?" The first thing that came to mind was a mental picture of my two year old great-granddaughter, Emily. My granddaughter went on to explain that the family was going on vacation for a week and, of course, that meant that Emily was going along. I was being asked to baby-sit Lily---my four month old great- grand dog! Lily is a Shih Tzu puppy! So, being a normal grandmother, I could never say no, so Lily has been spending the week with me!

I have had dogs with me my entire life, but never small dogs. It has been Border Collies, a couple Shepherds, and a very special Lab/Shepherd mix, along with numerous mixed strays that were always being dropped off near the farm. We always gave them a home and most turned out to be good and faithful companions. While they weren't all "big" dogs, none were ever as small as Lily. She is about the size of a small sack of sugar right now. So having a small dog like this is a new experience for me.

Lily is a bundle of energy and quick as a wink. She has acute hearing and the tiniest sound draws her immediate attention. She has a very keen sense of smell and one of the things that really gets her excited is the smell of marble cheese! She will hunt everywhere trying to find it. It didn't take her long to learn that if I am sitting at the kitchen table it usually means I have food! It is an excellent opportunity for her to become a spoiled great-grand dog!

They didn't bring any dog toys along so we had to create some. An ordinary washrag tied in a knot made Lily very happy. She can chew on it, carry it around, and loves to fetch it when I throw it for her. She has found it to be a lot more fun than one of my old slippers. When she gets tired her favorite place is to curl up on my lap in my recliner for a nap. That works really well, as I can get a quick nap also!

When she is sitting on my lap in the recliner, she is up where she can see everything on the television. She is sometimes very interested and watches for quite a while. When she sees something she doesn't like, you hear her start this low, throaty growl and then she barks loudly at the screen. That is exactly what I want to do when I try to watch some of these terrible shows that they are airing these days! When we are outside and she spies the horses she turns into a growling, ferocious watch dog and if you close your eyes you would think she is the size of a Rottweiler! The name Shih Tzu actually means "little lion" and that is exactly what Lily thinks she is!

The Shih Tzu is one of the fourteen oldest dog breeds. It is said that they were developed by Tibetan Monks and given as gifts to Chinese royalty. The purpose of a Shih Tzu is to be a companion and Lily follows me everywhere. My greatest concern is that I will accidentally step on her! My dogs were always a very important part of my life as a child and remain so to this day. I think every child should grow up with a dog. They listen to a person's troubles without judging or criticizing, they just give love! Lily will be Emily's special friend and companion as they grow up together. Having Lily here this week has been interesting, educational and a joy. I look forward to babysitting her again.

BORDER COLLIES HAVE ALWAYS BEEN MY BEST FRIENDS

I am a dog lover and always have been. As an only child growing up on a dairy farm, my dogs were my playmates and my companions. Stubby was the first dog I remember, a bob-tailed shepherd, who herded the cows. Later on there was Rex, a cocker spaniel-hound cross that was given to me as a puppy. We explored the pastures, finding all sorts of things, and followed my dad as he worked in the fields. I harnessed them together, hooked them up to a small wagon, and drove them everywhere. They were my team! I was broken hearted when Rex was hit and killed, while crossing the road at dusk on a summer evening, by a neighbor who had neither lights nor brakes on his old car.

On my eleventh birthday we drove to Creston, Ohio to purchase our first Border Collie. I have never forgotten when the door on a little red barn was opened and two adorable black and white puppies came tumbling out. I had to choose one and that was hard. His name was King and he rode all the way home on my lap. He was smart and knew much more about how a Border Collie should work than we did! He loved my Dad and went everywhere with him. My Mom, however, had a little problem with him. She had free range chickens and when King didn't have anything else to do, he would spend hours making every hen go back into the chicken house and when he finally had them all rounded up he would lay down at the door and keep them there. My mom would have to go get him so the hens could go back out!

Thus began a lifetime of owning Border Collies and what a joy they have been. Just as with dairy cows, the genetics in some have been better than others. Some have been so smart it was scary, others were not outstanding, but I loved them all. Our best were the ones bred by Jane Brugger at Winfield, Ohio.

Lady was one of the special ones, so smart and could figure out things that needed to happen without being told. She was very good at making the cows stay out of the fences. If she heard the fence creak she knew that somebody was grazing through the fence and she would rush to put them out. For almost a year after she died, if I saw cows eating through the fence, all I had to do was yell her name and they would back out!

My current Border Collie is officially registered as "Blue-Eyed Lass", as she has one blue eye and one brown eye. She came from Jane's and she is a smart one. We have to spell some things instead of saying them. We raised our calves in individual stalls and they had water buckets set in with them. If Lass heard them playing with the buckets and trying to upset them, she would rush to nip them and make them stop. We didn't dare say the word "bucket" or she would immediately look for the calf that was being bad. And if you wanted to swear at the cows, you better spell those words also, because when she heard cussing, she would

go and lay down on the milkhouse porch and refuse to work until a different person gave her a command! Lass is going on thirteen and over the years she has learned a lot of words! When our cows left, she was forced to retire, but she would still rather be a working dog.

I was recently very touched by a story that appeared in Farm World, a publication from Indiana. It was written by their Missouri correspondent, Matthew D. Ernst. The subject was "Haley", the first unregistered Border Collie to make the United States Border Collie Handlers' Association Cattledog Finals, placing ninth overall at the event in Leeton, Missouri on May 5, 2013. Haley was referred to during the trials as the "Cinderella Dog". Haley was found with a litter of pups and tied to a stop sign in Millersburg, Ohio nine years ago. It all started when Border Collie rescue headed by Wanda Heyman in Tiffin, Ohio, contacted Liz Muehlheim, a Cleveland physician, to foster Haley in 2004. Realizing that Haley needed to work, she took her to an Introduction to Herding Clinic held at Hado-Bar Farms in Nova, Ohio, and later she adopted Haley. Liz met and married George Muehlheim, a veteran Australian shepherd dog handler, and Liz was soon traveling to sheepdog trials with him. Haley has all the herding work she wants at their sheep and cattle farm in Ashland, Ohio. After Haley placed in the top ten, the city-girl-turned-working-dog-handler could not stop thinking about the dog's orphan origin. "And all I could think was that someone had thrown away a damn fine dog", said Liz. Congratulations to Liz and Haley!

MILKING A HIPPO

If you are asked the question, "How do you milk a hippopotamus?", the answer should be "very carefully!" I recently read a story in a local Sunday newspaper titled, "The Science Behind Milking a Hippo". The story was about Fiona, a new born baby Hippo born at the Cincinnati Zoo and Botanical Gardens. Fiona came in to the world six weeks early and was only about half as big as she should have been. She was too small and too weak to nurse from her mother, Bibi. Getting Fiona's strength up was critical and that meant she needed her mother's milk or colostrum, just like a baby calf. In order to get the colostrum they had to milk a 3,371 pound hippo---no easy task! Fortunately Bibi had been worked with for other reasons and was trained to go into a small closure where they gave her treats to encourage her to lean sideways so they could reach her udder. Reaching the udder was one of the biggest problems in trying to milk her. Fortunately the postdoctoral researcher at the zoo was raised on a farm and knew how to hand milk. Hippo udders are shaped more like pig ones. They have two teats. It is said that they are the only mammal that gives pink milk. The pink color comes from two types of acids that cause the hippo to release an oily secretion referred to as "blood sweat" from its body.

A sample of the hippo's milk was sent to Washington, D.C. to scientist Michael Power at the Smithsonian National Zoo. He runs the milk repository at the Smithsonian Conservation Biology Institute Conservation Ecology Center where there is a national milk bank with no fewer than 15,000 milk samples from at least 130 different mammal species. Until he got Bibi's samples, he had never seen hippo milk. It is high in protein, low in fat and sugar. Since Bibi did not have a lot of milk they went to work to prepare a formula for the baby hippo. Power's goal is to collect multiple milk samples from multiple females of a species. He wants to find out over time how milk changes and if there is differences between individuals of a species. In other words is gorilla milk gorilla milk or is each individual animal different. Giant anteater milk is the closest thing to hippo milk that he has seen.

I have hand milked a lot of cows in my time and I enjoyed doing it. To make hand milking a pleasure you needed a cow that didn't kick, an udder that was properly placed and positioned so you could get the bucket under it, teats that were the right size, a cow that milked easy and a sturdy stool. I learned to milk when I was four years old, milking a very patient and quiet Guernsey cow, one teat at a time, in a tin cup which I emptied into a bucket. Dad would finish her for me when he got done with the cows he was milking. It was our "social" time. We talked about the day's happenings, family and friends, the weather, etc. Dad used to count how many squirts it took to fill his bucket. It was always great to have that cow that needed a second bucket! Squirting milk in to the cat's mouth was always fun and some of our cats were well trained! It was family time together.

When we showed at the Ohio State Fair and milked our cows by hand, we always drew a crowd. There was always questions and someone always asked why there was soapsuds in the bucket. Cindy was featured on a Columbus television show milking her favorite cow, Jazz. I was once asked to house and hand milk a special show cow. Evidently her well known owner felt hand milking would be less stressful on her beautiful udder.

When I first put my cows on official AR (Advanced Registry) test, the tester was Ted White from Freeport, Ohio, in Harrison County and he could drop in anytime to test. You never knew when he might be coming. He not only tested all breeds in Ohio, he also traveled to other states as well. He was a "check tester" for cows that were making state and national milk and fat records. He could tell some very interesting stories about the check testing! Most of the cows that were making huge records for their breeders were housed in box stalls in deep, deep straw and were milked by hand. He never mentioned any names---just some of the things he found in the straw!

Just think what the price of milk would be if cows today had to be hand milked! Over production would not be a problem and milk processing companies would be visiting the farms and begging for the farmers milk! Growing up on a farm, you never know when the things you learn might come in handy! While I have never milked a hippo, I did body clip a six-legged Holstein heifer-----but that's another story!

SHARING WITH YOU A TALE ABOUT TAILS

In this day and age we shouldn't be surprised at the unusual things that thieves will choose to steal. However, I was taken back a bit when I recently read an article in Farm World. Thieves in southern Illinois entered the Southern Illinois Equestrian Center near Marion on two separate occasions and cut the tails on six horses. It is believed that the theft is tied to the market value of horsetails for use as extensions for show horses and for horsehair jewelry. The act of stealing horsetails is more widespread out West. Horsetail hair can sell for upwards of $80 to $100 each on the black market, especially hair with unique colors. There are websites for show horses that sell horsetail extensions for up to $500 each, depending on the colors and the weight. There is a growing market for horsehair bracelets that don't take much hair to make. Approximately three feet of the tails was removed from each horse. While it doesn't cause pain, it leaves the horses with problems during the summer months, as the tails are needed to swat away the biting flies. In Illinois the act of stealing a horsetail is a Class A misdemeanor that carries with it a maximum penalty of one year in jail and a $1000 fine.

Tails are not only important for the comfort of horses in the summer, they are also much needed by cows that go out to pasture. As I read about the loss of the horses' tails, it brought back memories of another tail.

Many years ago we were in need of a couple dairy cows to replace ones that we had culled. My Dad and I attended a public auction of 80 registered Holstein dairy cows near Navarre, Ohio. We arrived early and looked over the herd. We found only two cows in the entire herd that were the nice dairy type and the kind of udders we were looking for. We were lucky enough to bid those two in for $500 and $510 each, a pretty high price at that time. One of the cows was sired by an A.I. bull named Sir Wallace Design and had been bred in the herd of Kenneth Renner at Dalton, Ted Renner's father.

Her name was Sally and she had been housed with those other 79 cows in a straw shed type barn. We brought her home and put her in a stanchion barn. One of the things we noticed about her was that every night when my Dad went to the barn at 11:00 o'clock to check on things, she would be standing up eating while most of the other cows would be at rest. She wasn't a big cow, and we decided that those other 79 cows had been pushing her back from the feed and she wasn't going to let an opportunity to eat her fill go by! Her production increased and she was a very good cow!

Our cows were out on pasture in the summer, and as we watched them making their way to the barn one hot summer evening, we realized something was wrong with Sally. She was covered in blood from head to tail! As sometimes happens, she had been laying down and another cow had evidently stood on her tail. When she jumped up, the end of her tail and the switch were torn off. She had kept swinging her tail until there was blood everywhere. We hosed her down and got the bleeding stopped. We kept her in the barn for a few days until it healed and then let her go back out to pasture with the other cows. It was hot and there were lots of flies and the fly spray just didn't last on her, so she was miserable. I decided I had to do something to help her.

I cut a piece from a burlap feed bag, then cut it into strips of proper length, leaving a band at the top. I wrapped the band around her tail and I taped it fast. I had to experiment for a few times to get it just right. Once I did, Sally's "fake tail" would last for three or four days. By that time she would have the strips worn out, and I would make her another one. She was more content, her production didn't suffer, and I felt much better! During the cool months she didn't need it. Sally and her "fake tail" was with us for several years. It wasn't fancy and sometimes caused people to look twice, but it did the job!

CALL RENT-A-COW TO SAVE ON YOUR TAXES

As you drive down the streets in Florida you will see new buildings going up for business offices, medical complexes, condominiums, etc. and suddenly, on the same property you will be surprised to see a tiny herd of cows – maybe a half-dozen – in a wide fenced-in field of grass. These are the "Rent-a-Cows". My son and daughter-in-law saw the rent-a-cows on a recent trip to Florida to visit their sons for Easter. Some are just a few minutes away from Medical City, the University of Central Florida's sprawling campus of hospitals and teaching facilities.

It is known as Florida's greenbelt law. The statute is meant to preserve farmland by taxing it at a special, low rate. However some of the biggest beneficiaries have taken advantage of it by literally renting cows. The greenbelt law dates back to 1959 when suburban strip malls and subdivisions were being built. It was a troubled time for farmers, as land was assessed and taxed based on its most profitable and potential use and for the most part, that meant real estate. Citrus trees didn't offer the same returns as new condos. Farmers were forced to either sell their property or risk being priced out. The greenbelt law offered a solution by dropping rates for agricultural land.

The problem is in the vague wording. To qualify for the exemption, property owners are required to use their land for "bona fide" agricultural purposes. However the definition of bona fide is far from clear. Some developers offer their land free of charge, others actually pay the ranchers --- hence the loophole's nickname, "rent-a-cow". Some of the acts biggest beneficiaries are deep-pocketed developers who often take advantage of it by literally renting cows. The tax law allowed

Disney World to save $1.5 million dollars. Some of the tax breaks go to legitimate commercial farms, however most of the beneficiaries are not farmers. Residential properties don't qualify for the exemption. So you can't just rent-a-cow and put her on your lawn.

Florida is not alone. There are other states that have quirky tax breaks. In Alabama you can still deduct $1000 for building a radioactive fallout shelter. In Arkansas blind combat veterans may buy a car every two years tax free. In Hawaii residents can claim a $3000 deduction for taking care of "exceptional trees" on their property, as long as an expert deems them "exceptional". In New Mexico if you are over 100 years old and not claimed as a dependent, you don't have to pay income tax. In Kansas if a hot air balloon is tethered to the ground, passengers will be taxed for entertainment, but if the rope is untied, the ride is tax free. In Indiana marshmallow crème is tax exempt, but marshmallows are not. To encourage young men to marry, Missouri charges an annual tax of $1 to single men between the ages of 21 and 50. The law was passed in 1820.

HOLSTEINS AND BUTTER MAKES EVERYTHING BETTER

At last ---two days without rain! Today as I sit on my deck and write, it is a beautiful day with warm sunshine, a cool breeze, and puffy white clouds in the sky. I have three hummingbird feeders and the birds are coming to feed constantly. They flit, fight, and feed! It is so peaceful and tranquil. So sad that so many people all over this world never have the opportunity to enjoy the peace and beauty found in the rural countryside.

We had eleven inches (11) of rain in three days. Time and sunshine are needed to get things dried out! There is a lot of cleanup to be done in many areas. Hats off to the people who worked night and day to take care of flooded roads, downed trees, mud slides, cleared roads covered with rocks and debris, rescued people and more. It is pretty unusual to see a snow plow clearing the road in June!

Last Saturday I announced the Northeast Ohio District 1 & 2 Holstein show held at Lisbon, Ohio. What a great show! Eighty eight (88) registered Holsteins paraded the show ring. I will have a report on the champions for you later. There is a list of the winners and pictures on Dairy Agenda Today. The quality was outstanding --- beautiful cows with beautiful udders! The Holstein breeders in that area have accomplished so much in both type and production over the years! Seeing beautiful Holsteins, announcing the show and visiting with old friends and new ones is always a joy for me.

I have to talk about those great kids I see at every show. They range from "peewee showboys and showgirls" to age 21. They do a great job of showing their Holsteins. Hard working, responsible, confident, kind, respectful, the adjectives could go on and on! They take care of and work with their show cattle every day as well as helping with all the other farm chores.

Most are active in sports, school activities, and community happenings as well. These country kids are the best!

There was a small crowd on hand to just watch the show, including Holstein breeders who have been involved for years. One of those was Mrs. Marjorie Whiteleather, who will celebrate her 97th birthday in July. The Whiteleather family has been involved in registered Holsteins and members of the Ohio Holstein Association for many years. They have an outstanding herd of Holsteins and the Whiteleather prefix is well known.

As I have often said, I enjoy reading and do a lot of it. Over the years I have subscribed to many magazines of all types. Hardly a day goes by that my mailbox does not contain a letter encouraging me to subscribe to a magazine. Since so many of them contain very little except for advertisements the letters wind up in my bag of papers for my cousin's church to recycle. However, I always read them first. I did recently subscribe to one whose purpose was to promote healthy eating. There were many recipes using fresh produce and both red and white meat. However the majority of them called for ingredients such as kimchi, bok choy, tamari sauce, agave syrup, soba noodles, shisito, aleppo pepper, cointreau, orecchiette pasta and more. These ingredients are not available in our local grocery store! Therefore the magazine was not very helpful for my health!

This week I did receive an advertisement for a magazine that is a bit different. Included in the envelope was a brochure and the following caption quickly drew my attention, "Butter Makes Everything Better"! Hurrah! A magazine that encourages the use of a dairy product! My request for a subscription is on the way and my check is included. Any magazine that encourages the use of dairy products and supports farmers is going to get a subscription from me and I am looking forward to reading it!

HOW NOW BROWN COW, WHAT WOULD YOU LIKE FOR SUPPER?

I recently read an article titled "Study confirms corn can be replaced in dairy rations". A Cornell University field survey showing high-production dairies in northeastern U.S. and upper Midwest were successful feeding lower starch diets prompted interest by farmers in Northern New York to evaluate economically-feasible replacements for corn grain in dairy rations. This set me to thinking about some of the different things that cows owned by our family over the years had ate.

My grandpa and grandma on my Mom's side of the family were pretty poor when their family was young and growing up. They had a family cow, but couldn't afford to buy feed for her. So my Grandma cooked all her potato and apple peelings, and the parts of vegetables that couldn't be used in a pot on the back of the stove. She would mix some bran with it and that was what the cow ate. When I had my herd of cows, I made a mixture of mashed potatoes, sugar, and the potato water and fed it to any baby calf that had scours in place of milk. That always seemed to help heal their stomachs. I guess I thought of doing that because of what my grandma had to do to take care of her Jersey cow. Giving them yogurt is recommended now.

When I was young, we always pastured the cows in summer and the grain was limited. My Dad would grind corn, oats, some linseed meal, and some molasses in the "hammermill" for their ration. When fall came, we always had a lot of pumpkins planted. We would gather them and one of my jobs was to cut them into pieces and put them on top of the cow's feed. The cows always ate them and seemed to like them. I have no idea if it helped their production.

When our daughter, Cindy, was in 4-H and showing at state and district Holstein shows, she had a big white heifer calf sired by a Canadian bull that few people ever heard of, Langview Intensifier. Show calves were kept in the barn during the day and turned out at night in a grassy lot close to my parents' house. My Mom always threw her peelings and waste from fruits and vegetables over the fence into the lot. We soon discovered that the calf was eating just about everything that was thrown over the fence --- potato peelings, apple peelings, grapefruit and orange rinds, cabbage leaves, etc. Her nickname became "Garbage", "Garby" for short. She grew to be a big calf and an outstanding heifer, and went on to win All-Ohio honors for Cindy. People all over Ohio knew who "Garbage" was!

How much her "diet" helped her to grow, we don't know, but as an adult cow one of her favorite things were over-ripe cucumbers!

Many years ago there was a bull named Bill Bess Burke and I doubt very much that there is anyone out there who remembers him but me! "Burke" bloodlines were very popular at that time---- Pabst Burke Dell, Wis Burke Ideal, Raven Burke Ideal, and more. We were milking a two year old by Bill Bess Burke and her name, of course, was

"Burky". In the fall after the hay was all made and the crops were all off the fields, some electric fence was put up around the barn and the cows were pastured in the hay and corn fields. Included in that area were two big gardens. Since we had gathered and canned everything we wanted from them, it didn't matter that the cows could run over them. Every morning when the cows were turned in, Burky would run to the garden and eat her fill of left-over tomatoes. It was so funny to see her squishing them in her mouth and the red tomato juice flying everywhere! She completely cleaned up the tomatoes, red and green. Why she took a liking to them, we never knew!

I have read of herds, especially one in New York, that hauled all the stale and outdated bread and sweet rolls from a nearby bakery and mixed them into their ration for their dairy cows. As I read articles about rations, I wonder how many different things have been tried in rations for dairy cattle. So much money and effort is put into studying and fine tuning rations for all our animals these days, not just dairy cows. Sometimes I wonder how our ancestors kept their livestock alive and well without the "experts" to guide them. Maybe they just used some "common sense" and whatever they had to work with!

COW CUDDLING COULD BE A BEGINNING

A turkey that wasn't done in time for dinner, corn bake that was done a little too much, a one year old who found the bathroom door open, water in the commode and a roll of toilet tissue, family members who were late, a hot water heater that quit working before the dishes were done, children running, laughing and chasing each other everywhere, baby's crying and a noisy football game. In spite of all that it was a Happy Thanksgiving!

This time of year brings on anxiety, frustration, short tempers and lack of sleep. I recently ran across an article on a type of therapy to deal with all those problems --- "cow cuddling". There is a farm in another state that offers cow cuddling, as well as numerous other activities for therapy. The service is offered to an individual or to couples. An appointment is made and the length of time chosen. The customer can pet, brush, play with, talk to or just sit with the cow for the chosen amount of time. The cost is $300 for an individual for 90 minutes or $375 for a couple. If you spend time just looking at the cow the cost is $100 for up to two people and $175 for four people for 45 minutes. This is a new wellness trend to relieve stress and connect with nature. The San Francisco International Airport features an adorable therapy pig on hand for stressed or anxious travelers.

If only Mama-Red was here today! She would have been the perfect cow for cuddling! She was so quiet and "laid back" from the first moment I saw her at the Ohio Holstein Sale at the Richland County Fairgrounds years ago. I fell in love with her and was the last one to raise my hand. Although due to calve at any moment, and a trip home in the middle of the night with fog so thick you couldn't see a thing, she calmly walked off the trailer, walked into her new box stall and went to munching hay. Harold and Marilyn Hoffer were not so calm, as the drive had been a terrible one, and they still had to back track their route for several miles, as they had completely missed the driveway to a farm where they were to deliver another cow! As the years went by Mama-Red continued her quiet ways. When the green chop wagon pulled into the pasture the other cows would fight and hassle to get in to eat. She would just patiently stand back and wait until the crowd cleared out and then walk in and get her fill. She was often seen in the middle of the night eating at the wagon while the rest of the herd lay chewing their cuds. My granddaughters when small would sit on a bucket or stool in front of her and read books or sing to her as she rolled her eyes and kept chewing. When school children came to visit she was the one they tried to hand milk and she was always quiet and patient as they squeezed away. We often found the bucket sitting under her after the children had moved on to other activities. She lived for 18 years and would have been perfect for "cow cuddling"!

At this time it is impossible for dairy farmers to support their farms and families and keep bills paid on the income from selling milk alone. There is a need to find other ways to bring in income and dairy farmers must come up with new ideas and ways to do that. For a fee, what about offering "educational opportunities" for people to learn about dairy farming. There are so many people in our world today that know nothing about cows and farming. Many have been given only misinformation about what happens on a farm. Individuals could be offered the opportunity to be in a milk parlor and learn what happens there and how it is done. Groups could sign up and for a fee observe the planting or harvesting of crops, what their purpose is, how weather affects them, why sprays need to be used and more. Organic farms could hold sessions to explain how they do things and why they think their products are better for consumers. All sessions would have a farmer or leader to explain everything and to answer questions. Special sessions could be held for children to play with and learn about how baby calves, as well as other baby farm animals, are born and cared for. Farm families who show their animals could schedule sessions for children to help give a spring calf a bath or to learn how to handle animals in the show ring. Parents are quite willing to pay for their children to participate in numerous other types of activities. Not only would consumers and their families be properly educated about dairy farming, the care of animals, and growing crops, but the farmer would have an extra source of income.

Not only would opportunities like these be beneficial for the public but also for the farmer. Therapy on the farm for people could ease their anxieties and frustration. The fees for making the sessions available could ease the anxieties and frustration of the farmer. Cow cuddling could just be the beginning of happenings in the future on the farm!

REMEMBERING HOPALONG CASSIDY
AND OUR FIRST HOLSTEIN

Ok you can stop chuckling now. Yes, that is me in the picture, wearing my Hopalong Cassidy hat and holding those Guernsey calves. For any of you "younger" people who never heard of Hopalong Cassidy, he was a popular hero in Western movies. He wore a black outfit, a black hat, and rode a white horse named "Topper". He was a fictional cowboy created in 1904 and was portrayed in the movies by William Boyd. He only drank sarsaparilla, never anything alcoholic!

On Saturday nights the milking was done early and then Dad, Mom, and I would be on our way to Carrollton. Everybody went to town on Saturday night! The streets would be filled with cars and if you weren't early you had a hard time finding a place to park. People would be lined up on the sidewalks talking, the men about crops, the weather, livestock, and the women usually gossiping. The first stop would be the grocery store to get needed supplies, sugar, flour, yeast, matches, salt,-- the basic needs, and if I was lucky, a poke of candy. Then it was off to the local movie theater for the late movie. There was always an early showing of the movie, then a news reel and a showing of the titles of next week's movies, and then the same movie would run again. On Saturday nights the movies were usually Westerns, with stars such as Roy Rogers, Gene Autry, Tim Holt, Randolph Scott, Rex Allen, Hopalong Cassidy, and one of my favorites, Lash Larue. He dressed in black, wore a white hat, rode a beautiful big black horse, and corralled the bad guys with a huge blacksnake whip. I always wished I could learn how to crack one of those whips! Of course, Hoppy was a favorite, hence the hat, and I toted one of his gun sets for awhile!

Our first Holstein was a bull calf. Dad bought him from a dairyman, who had a well known herd of good Holsteins. We brought him home in the trunk of our Chrysler coupe with Dad holding him, Mom driving, and me being thrilled to be bringing home a full blooded Holstein. There were no registration papers. At that time we didn't even know what a registration paper was. We didn't know a lot about Holsteins. We just knew that Guernsey and Jersey breeders were constantly saying, "If you drop a quarter in the bottom of a full pail of Holstein milk, you can still read In God We Trust on it!

Our herd was such a mixture of Guernseys, Milking Shorthorns, Brown Swiss, and even a couple of half Hereford crosses, we could hardly wait until we had some cows with that Holstein blood milking in our barn. Eventually Dad heard about "artificial insemination"

and signed up to have our cows bred by the COBA technician. At that time the technician would start out the day with semen from two or three bulls and if he came late in the day to breed a cow, he might only have semen left from one of them. So you didn't always have a choice. We bought our first Holstein cow at a farm auction. She was sold as being registered, however we never received the paper. She was a very nice cow with a beautiful udder and she gave us some really good grade heifers. Changes were on the way, and after purchasing my first Registered Holstein in 1953, we would eventually reach our goal of owning a complete herd of Registered Holsteins.

Like the movies, the breeding of Holsteins has constantly been changing. Today we have numerous AI studs in the business of selling semen to breed our cows. We have the option of collecting young bulls of our own and using them. Semen is available from other countries. We can use sexed semen. The opportunity to flush top cows is available to everyone. In vitro-fertilization is becoming popular. There are a multitude of bulls and bloodlines we can choose from. Holstein breeders have learned how to breed their own cows and implant the embryos. We have records and information of all kinds to help us do a better job in mating our cows. Genomic testing and information is available to us. There are classifiers to help us learn more about our cows, and mating specialists to help make breeding decisions. We have learned how to better feed and care for our cows and in return, they give two to three times, or more, the pounds of milk they gave in the past. There is so much information and knowledge out there, all we have to do is use it!

I have no idea what ever became of my Hopalong Cassidy hat, but Western movies kept improving, and I still enjoy them. Today's Westerns are nothing like those old time movies, and they are no longer just in black and white. Just like the movies, today's Holsteins have kept improving and they are no longer just in black and white either!

FARMERS CAN CONTRIBUTE TO HIGH FASHION

We have all been aware of the recent fashion trend of wearing "distressed" jeans. High prices are paid for jeans that are ripped, frayed and have holes in them. Numerous TV actors and movie stars show up on talk shows wearing them and especially seem to like the holes in the knees. People in the general public are quite willing to pay top dollar for the brand new ripped and distressed jeans. It seems the newest fashion is "dirty jeans"--- jeans that look like they have dirt on them! They are jeans that look like they have been worn by someone with a dirty job ---made for people who don't!

Nordstrom is an American chain of department stores headquartered in Seattle, Washington. Nordstrom offers "Barracuda Straight-Leg Jeans", which are heavily distressed medium blue jeans with crackled, caked on muddy coating at a cost of $425 a pair. They make sure the mud is on the knees, pockets, and crotch of the jeans. There are other jeans available at higher prices and it isn't even real dirt!

Now we all know that when it comes to dirty jeans nobody does it better than farmers! These dirty jeans were the subject of a recent newscast where they took brown dirt from a pot of flowers and rubbed it all over the jeans they were wearing. We all know that dirt comes in different colors – brown, black, red clay, as well as the different colors from other things besides soil. When cattle are on pasture the manure you get on your jeans has a green cast to it. A calf with the scours can give you some yellow stain. When a farmer receives an injury and bleeds he always wipes the red blood on his jeans. There are various shades from grease, oil, rust and the many more ways that farmers get their jeans dirty.

Selling their dirty jeans could be a source of extra income for farmers. Farmers and their families come in all different sizes from all over this country, so finding jeans that fit shouldn't be a problem. Wives and mothers would be happy, as the loads of laundry needing washed every week would be greatly reduced. Money would be saved as less detergent and fabric softener would be needed. Gals would have more time to spend helping in the barns and fields where they could get their own jeans stained and dirty and sell them to other gals for their own extra income. The kids would love the idea of being allowed to play in mud puddles and other places to get their clothes dirty! People could order the "deluxe style" jeans which would give them the choice of manure stains from any of the different farm animals or certain colors of dirt. Farmers' jeans are "authentic", not reproductions, so should be more valuable. Add some baler twine for a belt, a smokeless tobacco can in one back pocket and a red bandana handkerchief in the other, pull on some "Tingley" boots and you have high fashion! How long before we see it on the "red carpet"! We have already read articles telling us that our jeans should not be washed. Farmers' dirty jeans are an ideal product and farmers definitely know how to "manufacture" them!

Perhaps Dairy Agenda Today could provide a site for farmers to advertise their dirty jeans. Wouldn't it be great if Brad Paisley wrote a song to help the farmers sell them? How long before those high fashion people are going to want that "farm fragrance" to go along with those special outfits? There is nothing like the smell of newly mowed hay, silage coming down the silo, the smells in the milking parlor, and more. You just can't find those in the big cities! There is no doubt in my mind that farmers would be quite willing to provide whatever those people in high fashion desire and at the prices they are willing to pay. Nordstrom also offered a "medium leather wrapped stone"--- a medium sized stone wrapped in leather with special stitching--- "sure to draw attention wherever it rests" read the description. The cost ---$85. The product sold out last year!!

JUST THINKIN' ABOUT DRINKIN'

I am acquainted with a young man who has an important job with a very well- known organization in another state. Along with numerous activities, he handles finances and often oversees orders for various items. His organization gives out numerous Christmas gifts and before this Christmas he was requested to order bottles of a very special type of whiskey to be given as gifts. This whiskey comes from a special brewery that brews and sells a limited amount each year. It is considered very special because it is made from a sweeter mix of corn, wheat and barley malt, handcrafted in small batches and aged in new charred oak barrels. It is expensive because it is rare and amounts are limited. People will go their entire lives without an opportunity to sample it. Finding a bottle is so rare that the government in the state of Ohio holds a lottery to see who gets a bottle. The prices can vary, depending on who has it. When the young man finally found some for sale, he ordered twelve (12) bottles for his organization to give as Christmas presents at a cost of $3800 per bottle. Do the math!

I was just thinkin'--- what if the milk from the supreme champion at World Dairy Expo, Rosiers Blexy Goldwyn, was put in specially designed bottles with her picture on them and offered for sale? Are there rich people out there who would like to dunk their chocolate chip cookies in a glass of her milk or pour her milk on their corn flakes at breakfast? Would they pay extra to give their children milk to drink from an elite, prize winning cow? Would sports stars who pamper their bodies and make milk a part of their diets pay more for milk from prize winners? What about the rich people in other countries who like to spend huge amounts for items from our country?

This idea wouldn't be limited to supreme champions only. Every breed that shows at World Dairy Expo has a grand champion. Milk and dairy products made exclusively from each of those champions could be offered. Would butter made from the milk of the grand champion Jersey sell for more? What about Swiss cheese from the Brown Swiss champion or special dairy products from the A2 champion Guernsey? There are numerous ideas that could make milk and dairy products from champions appeal to the public. They could also be incorporated into many other products. What if Rosiers Blexy Goldwyn was the star featured on the Wheaties box and her milk was available to pour on those Wheaties?

There are dairy champions selected at state shows and many other shows in numerous states including Ohio. What if some of these champions were housed, cared for and milked at a special dairy. The milk could be bottled and marketed as "Milk from Ohio Champions". All breeds could be included or the breeds could be kept separate and the milk and dairy products labeled as being from the specific breed. Breeders in any area could use this idea and sell dairy products from a special herd of champions.

Would people pay more for dairy products from champions? Isn't it obvious that there are people out there who are willing to pay whatever it takes to get exactly what they want? So why not offer them some "special" dairy products. If Rosiers Blexy Goldwyn gives eighty six (86) pounds of milk a day, that is ten (10) gallons or forty (40) quarts a day. If each quart of that milk was sold for the same price as a bottle of that special whiskey --- DO THE MATH!! Just thinkin' about what people are drinkin"!

SOUP FOR THE DAIRYMAN'S SOUL

It was a long night, freezing rain, then snow, blowing winds, drifts everywhere. Morning brought colder temperatures, tractors that didn't want to start, frozen water pipes, plowing snow to get the milk truck in, a very cold newborn calf, and more! Finally after several hours and a lot of hard work, things are back in order, everything seems to be working, and that new baby heifer calf is tucked in and warm. She just might be a winner this summer!

As you enter the mud room or basement to get rid of the bulky coat, Carharts, boots, toboggan, and gloves, you are greeted by a beautiful aroma and it isn't coming from your boots! You make your way to the kitchen and there, simmering on the stove, is just what your body and soul needed --- hot soup. It is the time of year when thoughts turn to soup, and farm kitchens all over the country have soup on the stove or waiting to be heated in the microwave.

When you check out soups you find that there are so many different kinds. Not only are there recipes for the ones we have made over the years, potato, chicken noodle, vegetable, chili, etc., there are so many other nourishing ones containing numerous ingredients. You can find soup recipes from all over the world! I found it interesting that so many of the soups are made with some type of dairy product --- milk, cream, butter, sour cream and cheese of all kinds. Recipes for both apple soup and sour cherry soup use dairy products. Perhaps we should be doing more advertising to encourage those types of soups for consumers.

I can't think of anything better for this time of year and those cold, snowy days. I find soup to be nourishing, comforting and a stress reliever. There is something about sitting down to a hot, steamy bowl of soup and as you slowly savor each spoonful not only are you warmed and nourished, it also seems to ease and relieve the stress you are under. Several years ago there was rain and wind on Christmas Eve that turned into a terrible ice storm overnight. On Christmas morning everything was covered in ice, fences, roads, tree branches were broken down and electric wires were down. In order to make our way to the barn we had to put the spikes that railroaders used in icy weather over our boots and hold to the fences, gates and whatever else we could hang on to. It was three o'clock in the afternoon before my family could leave things at the barn and go to the house. There was no Christmas dinner to enjoy, everyone had to help get things under control at the barn. Fortunately there was a pot of leftover vegetable soup that had been made two days before. That was our Christmas dinner and probably the best tasting vegetable soup that any of us had ever tasted! We celebrated Christmas and opened presents two days later.

    The simple things in life can help us get through the hardest challenges. As dairy farmers begin to make their way through this new year of 2018 they are faced with many difficult problems and challenges. While it won't solve the problems, there are times when just a simple bowl of hot soup can help us to get through the day and survive. Take time on these cold winter days to enjoy your favorite kind! Don't forget the crackers!

DAIRY FARMERS PLAY THE GAMES

The one thing a dairy farmer never has time for is watching television game shows. Yet every day they participate in games like we see on television. Men and women rise every morning to spin the "Wheel of Fortune" in dairy farming. There are no guarantees as to where the arrow will stop. It could be good spins throughout the day with the weather co-operating, the machinery working, and all going well with the dairy herd or it could be "lose a turn" at having a good day with a visit from the milk inspector or numerous problems to be dealt with.

Dairy farmers have to participate with Mother Nature in the "Amazing Race" to get crops planted, grown, and harvested. Will there be too much rain or drought conditions, lots of warm sunshine or late frosts, will the harvesting be done before the snow falls? It is a never ending race and the farmers have to hope they can make it to the finish line and have a good supply of feed for their livestock.

Dairy farmers play "What's My Line" as they decide which breed of dairy cattle they want to raise and work with. There are many factors that come in to play --- color, size, production, cost, the ability of the chosen breed to handle the environment they are in, or should they just go along with the breed that they were raised with and know.

Dairy farmers play "The Match Game" every day no matter which breed of dairy cattle they have chosen. They must select the bulls to be used in breeding or flushing their cows. The goal is to have outstanding calves to raise and improvement in each generation. There are so many choices --- young or old bulls --- proven or unproven --- genomics or not --- all with outstanding pedigrees and information behind them. No matter how much research they do or information they study, there is that "bit of luck" that is still involved. If they are lucky enough to breed that very special heifer or cow, you may see them "Dancing with the Stars" at World Dairy Expo!

At times dairy farmers come up against "The Wall" and there are no balls to drop for needed cash. So they must visit their bankers and lenders and suggest "Let's Make a Deal" so they can continue with their dairy operations. Making the decisions there can be very difficult for all players involved. Will the banker put enough money in the "briefcase" and will the farmer make the right choice that allows him to stay in the game?

Dairy farmers are in "Jeopardy". The category is dairy farming. The answer is --- no one knows --- the question --- how many dairy farmers will go out of business in the next year? Every dairy farmer I know wants to be a "Survivor". They pray that their role as a dairy farmer won't be cancelled. When asked "Who Wants To Be a Millionaire" most dairy farmers will answer, "To Tell The Truth" we just want to make a decent living so we can own our farms, pay our bills, and have a comfortable family life. If the "Price Is Right" we can do that, but currently the price we receive for our milk is all wrong". Every dairy farmer just wants to enjoy "The Game of Life" on the farm and keep on playing.

LOOKING BEYOND A DIFFICULT DECISION

I recently heard about a comment from a dairyman saying that he was "tired of hearing about all the gloom and doom in the dairy business and about dairymen having to sell their cows". I decided that as a writer perhaps I should look for the "up" side of not milking cows and write about some of the good things that can happen after the cows are gone.

You are no longer committed to milking cows two or three times a day. You don't have to get out of bed at four o'clock in the morning or go to bed at nine o'clock at night. You can stay up to watch the final score in the Super Bowl, that John Wayne western movie or any other late night television program that you enjoy. There is that enjoyable feeling when you automatically wake up early in the morning (as most dairy farmers do) and you can just roll over and go back to sleep! You no longer have to stay up all night to look after a cow that is sick or calving, trying to catch forty winks on some bales of hay or straw while three cats are purring in your ear.

You no longer have to put up with milk inspectors (especially the ones who seem to enjoy their position of power} constantly nit-picking about every little thing or worry about not passing those state inspections. No worries about bacteria or somatic cell counts. If you want to, now you can choose to paint your milk house and milking parlor green with purple polka dots instead of white. Green always did do a better job of hiding the splatters! And what if old Shep did sneak into the milk house and was lapping up the few drips caused by the seal that needed replaced on the milk tank, he wasn't going to bite!

No more having to fix broken gutter cleaners that always break down when the gutter is full. No more trying to find a dealer on a Saturday or Sunday who will open up to get you the parts needed to fix it. No more worrying about getting the tractor started after a big snowstorm so you can plow out the driveway to get the milk truck in. No thawing of drinking cups or water lines following a week of far below zero temperatures. No barn clothes and boots sitting in the mud room or laundry room permeating the entire house with their fragrant scent. Now you can wear a new pair of jeans without automatically bleaching them with Clorox splashes.

    There will be more time for other activities. You can attend ball games and other sports events, tractor pulls, 4-H and F.F.A. happenings, graduations, go fishing, just sit on the porch swing, take a walk in the woods or go for ice cream with your children or grandchildren. You can take the time to visit elderly relatives or a lonely retired farmer that you have known all your life. Now you can become more involved with community activities, serve as an officer or director for your local breed organization, help with youth programs. Help a youth with a project calf, share your experience and knowledge and follow the progress made. You can attend events that happen between four o'clock and nine o'clock in the evening. You can read a good book. You can take a vacation without worrying about what is happening at home or phone calls about how to treat a special cow in a crisis.

    Think of the bills you will no longer have to be responsible for! Veterinary bills, medicine, trucking, semen costs, milk house supplies, paper towels, Clorox, teat dip, inflations, hoof trimming, boots, feed, etc. However, you will probably have to start buying cat food.

    You will be able to actually sit down and enjoy meals with the family. You can enjoy a real breakfast with bacon, eggs, toast and a couple glasses of milk instead of grabbing a cup of coffee and a donut and rushing to get the milking started. You can sit down in a favorite restaurant and eat a decent lunch instead of going through the drive-thru for a sandwich, French fries and a soft drink because problems came up at milking time and you had to run to town for parts and now you are behind! At supper time you can take time to talk to your spouse about the day's happenings and ask the kids about school and their activities instead of coming in late to a cold supper and the kids already in bed. You can take your spouse or that special person on a "date" to a favorite restaurant or a special party.

    No one ever promised that life on a dairy farm would be easy. Dairymen and dairy women constantly face problems of all types --- rules and regulations, pain and grief, birth and death, disappointments and hardships. Anyone who chooses dairy farming as their occupation has to love cows to stay with it. To have to choose between selling their cows or trying to get through these turbulent times has to be one of the hardest decisions they will ever have to make. Dairy farmers have more strength, courage, and faith than anyone else I know. They will find their pathway through these difficult times. Even on the darkest day there can be a ray of sunshine.

YESTERDAY WE LOST ANOTHER ONE

Yesterday I sat and watched as a herd of ninety seven (97) registered Holsteins sold at auction. The herd was owned by a young dairyman in my local area. He is married and they have a young daughter. I arrived early and had a chance to talk with him before things got so busy. When I asked him how he was doing I could see the glint of tears in his eyes. This young dairyman had taken over the herd from his father, who wanted to slow down. He has always been involved with the cows, helping as soon as he was old enough. He could recite the pedigree of every cow in the barn. He was involved in 4-H and district shows, taking one of their great cows all the way to World Dairy Expo. It was obvious that he loved his cows and this would be a tough day for him.

I walked down through the barn and snapped a few pictures of this nice group of cows. The cows were well taken care of, good udders, lots of type, good feet and legs, an outstanding group of red and whites, as well as the black and whites. People were reading their catalogs and looking them over.

The sale was held in a tent and it was filled with dairymen, their families, some very well- known Holstein breeders, friends and neighbors. The auctioneer was a cousin, the auction crew consisted of relatives, paid workers, and friends. A relative would welcome the crowd and talk about the family and the farm. The young dairyman remained in the barn. Talking that day just wasn't easy for him to do.

The sale began, the cattle looked good in the ring, bidders were waving their hands and nodding their heads, bids were coming in on the internet to Cowbuyer. The auctioneer kept things moving along, as one by one the Holsteins were ushered into the ring and the pedigree reader talked about each one. And then it was over, the last animal exited the ring as the hammer fell and the buyers headed for the auction trailer to pay for their purchases, get loaded up and head home. The large tent was soon empty. You could hear the hum of the equipment as cows were being milked out in this dairyman's barn one last time.

This was a young dairyman who grew up with cows and learned to love them. Sadly due to current conditions that dairy farmers face today, he had to make the decision to sell. Monday morning he will start a new job with a farm equipment company. His life will change. There will be a regular pay check, eight hour work days, weekends off, time to attend his daughter's activities, less stress and more family time. The barn will be empty, the milkers will quietly hang in the milk house. There will be no sounds from cows wanting milked and hungry calves. All will be still.

The consumer will walk into his local grocery store and purchase a gallon of milk. They expect it to be there. He or she has no idea what the life of the dairy farmer who made that milk possible is like. The thought never crosses their mind. Do they ever think about the fact that without the dairy farmer there would be no milk to drink, put on their cereal or cook with, no ice cream to enjoy, no yogurt, no cottage cheese or cheese of any kind, no butter, no dairy treats for their kids? Do they ever stop to think how necessary not only dairy farmers, but all farmers are in their lives? They have no idea how a dairy farmer feels when he has to sell his cows.

Our government officials never seem to realize how important farmers are to this country. They just keep enacting laws and regulations that make it more difficult for all farmers to farm and allow more imports to come in. Don't they realize that farmers don't want hand-outs from the government? Farmers just want to make a decent living that will allow them to provide for their families, pay for their farms, and improve their lives. Everyone in our country has to eat to survive. Do our government officials ever think about where the food they enjoy comes from?

Our farmers would like to be able to turn their farms over to children or family members and have their family farms continue in production for years and years. They would like to see young families who want to be farmers able to fulfill their dreams. Instead we are seeing dairy farmers giving up and getting out all over the country. Yesterday we lost another one.

"SUMMERTIME AND THE LIVIN' IS EASY"

Those are words from a popular song written several years ago. They are much more appropriate now than they were back then. It is summertime and hot, with ninety- plus degree weather, but these days we have ice makers, refrigerators, fans, and air conditioning everywhere, in homes, offices, stores, even in barns. That was not the case when I was growing up, and I am sure that people look back to those days and wonder how our ancestors, and especially the farmers, made it through the hot summers.

Not everyone had an electric stove, as they were a rare and high priced item. Cooking was done on coal stoves. Some farm wives were lucky and had "summer kitchens", which were small buildings away from the main house, where all the meals were cooked and served, thus keeping the main house cooler for sleeping. Everyone drank lots of water, lemonade, and iced tea and ate lots of fresh vegetables from the gardens. The canning of meat, fruits, and vegetables in preparation for winter was also done there.

Sleeping could be difficult in the summer heat. I can remember my Dad taking some old blankets out and putting them down on the ground under the big maple tree and we slept there a few nights. The ground was hard and there were the occasional bugs, but I thought it was fun, and it was the closest I ever got to going camping!

The cows were out on pasture day and night. They would eat early in the morning and then head for the "alders" along the creek in the pasture bottom. It was amazing how cool it was there all day long. They would head home for milking around five o'clock. Milking time could be pretty hot in the evening, leaning against the warm flanks of the cows. Once they were done, they would fill up at the watering trough and head back out to graze as the evening cooled down.

The pigs were kept in fenced in lots, and of course they had their "wallowing holes" filled with wet, cool mud. They spent most of their time there, coming out only when it was time to "slop" them. The days were hard on the horses, as they had to keep working. Lots of cool water and giving them time to rest under a shade tree kept them going, but they were glad to get back to the barn for their oats and to be relieved of the heavy, hot harness and allowed to cool down and rest for the next day.

Carrying water to the fields for the men who were bindering and shocking grain or making hay was a chore for us kids. Some people didn't have refrigerators to make ice, so we would fill the jugs from the cold water that ran in the spring houses. Some of the fields were far away, and we made a lot of trips. We would have several days in a row when it was in the nineties and we made hay every day. It would be loaded on the wagon loose with a hay loader, quite a modern improvement from raking it into huge piles with a "dump rake" and then pitching it onto wagons by hand with a pitchfork. My Dad often talked about the time my Grandpa Wagner was almost killed in a dump rake. He was raking hay with a team of young mares that were only "green broke". Something spooked them and they ran away. My grandpa fell off the seat and was caught up in the dump part of the rake that gathered the hay and was dragged along as they ran. To this day no one knows why the dump handle suddenly released by itself and dumped him out. That was the only time it ever dumped without someone pulling the handle!

Wagon loads of hay were taken to the barn where "hay forks" were set in the hay and horses were used to pull it up to where it latched onto a track and went to the "mow", where men with pitchforks placed it around so the mow could be properly filled. It was a hot job! Some people also filled a mow with straw at threshing time, another hot, dirty job! The men would come out of the mow covered with dust and wringing wet with sweat.

The outfits for farmers in those days were straw hats, short sleeved shirts,(some wore light weight long-sleeved shirts), trousers, heavy work shoes or boots, and large bandanas to wipe the sweat. Shorts were very rare and then only occasionally seen on children. If they didn't have corn to hoe or other chores to do, the kids would head for the creek to wade in the cool water, catch crawfish, and just enjoy nature. Some were lucky enough to have a "swimming hole" located near them.

Once again it is summertime but the livin' is much easier! With the weatherman predicting temperatures in the nineties again this week, it is nice to know we can turn on our fans and air conditioners and keep cool. Right now, I am going to go get a glass of ice cold lemonade!

GOING AROUND IN CIRCLES
CAN BE VERY BENEFICIAL

There are times when I just feel the need to go around in circles! When that happens, I hop on my John Deere and away I go. I can just enjoy the sweet smell of newly mown grass and listen to the peaceful hum of the motor. No need to make any decisions, telemarketers can't find me, around and around I go, with all sorts of ideas and thoughts wandering through my mind. It can be relaxing.

As a child, I loved to wander through the pasture fields, following all the cow paths, my faithful pal "Stubby" at my side. I would sit along the creek bank and toss pebbles into our "swimming hole", and watch as the circles spread out on the water from the impact. What a jouous and peaceful time.

As I grew older, I loved the days when my Dad would allow me to play "hookey" from school to work the fields for planting. With Dan and Prince hitched to the cultipacker and drag, around the fields we went, listening to the singing of the birds, the jingle of the harness, dreaming of the future (maybe  someday breeding an Excellent cow), smelling the fresh turned dirt, and enjoying the feel of the warm sun on my back as we circled the fields. In later years it was an Oliver 77, and the harrow. I took such pride in getting that finishing furrow smoothed out perfectly, just the way my Dad wanted it.

I like to think of not only the Ohio Holstein Association and all breed organizations as special "circles". Each is a circle of special people with one thing in common, the registered cow. Our love for her and our joy in her brings us together to form a circle with so many ripples that can touch us in so many ways.

There are so many people to get to know and friendships that form and last forever. There are District clubs to belong to and most need and welcome new members to take part in their projects and bring them new ideas.

WORDS USED THE MOST IN RECENT DAYS ---"IT IS HOT!"

It's Monday, a load of laundry is washing, another load is drying, the sweeper needs to be run, the telephone is ringing (mostly telemarketers) my column is due, and IT IS HOT! Those were probably the three most used words in the English language last week, as everyone suffered through the intense heat wave. As thermometers hit the 100's mark one could not help but wonder how our ancestors ever survived without air conditioning! But they did, going about their work every day, as it had to be done, in spite of the heat. How did they do it?

Farmers had no choice. They had to get hay and feed put in for the winter for their livestock. Threshing days could be brutal! The sun beat down as wagons were loaded with grain and hauled to the threshing machine. Usually threshing was done outside, with the straw blown into straw stacks, but some farms blew the straw into a mow and machines were set up in the barn where there was limited air movement. Some of the men went into the mows to move and stomp down the straw, and when they came out they would be black from the dust and weeds. Wheat and oats field weren't sprayed for weeds in those days, so the straw wasn't as clean as much of the straw you see today. The grain ran into burlap or special cloth grain bags and was then carried by a line of men to the grainerys, where it would be dumped into huge bins. It was hard work, and their clothes would be drenched with sweat. It took tough farmers to work at that all day.

Huge meals were prepared for their lunches, and usually on coal stoves. The women had to contend with the heat, as they baked and prepared to feed everyone. Some farms had "summer kitchens". They were out buildings where they cooked, so that the main houses didn't get so hot. My Mom didn't have that luxury.

When we threshed, not all of the help could sit down at the table at one time, so they would all wash up in the big rinse tub that had been set out for them, and then some of them would stretch out under the huge maple tree in our yard until it was their turn to eat. It was such a cool place and there usually was a breeze there. They always hated to have to leave it and go back to work. I can remember several summer nights when my Dad would throw a couple quilts on the ground and we would sleep under that maple tree. In those days it seemed that every farmer wore a straw hat. In spite of the warnings from the doctors, you don't see them worn as much today. They always carried water jugs to the fields, and if they ran dry, we kids were sent to the house for a refill. Sometimes we made several trips.

Our dairy cows always knew where to go to beat the heat. They would pasture early in the morning and then head for the "alders", as we called them. They are large bushes that grow all along our creek banks down in the bottom of the pasture. It always was, and still is, the coolest place on the farm. Our cows could always be found there on the hottest days, standing knee deep in the creek or lying in the grassy spots along the creek bank. In later years when we green chopped for them, we always put the wagon as close to the alders as possible so they would be sure to eat.

In those days, Mother Nature made it hot for us but she still provided ways to beat the heat. Most farms had a pond or a deep spot in the creek where you could go swimming, and there were lots of shade trees. Big cities had parks, trees, and swimming pools. Perhaps it helped that there wasn't so many high rise buildings to block the air flow. And they didn't have blocks of concrete and miles of asphalt pavement. The news recently showed an egg and pizza being cooked on the hot pavement during this heat wave. I don't remember being able to cook an egg on the dirt road. There was lots of dust! But when it rained, the dirt absorbed the water and the earth cooled. These days it just runs off the blacktop and causes flooding! What is that saying? "It's not nice to fool Mother Nature". Perhaps she doesn't appreciate all the black top, cement, and buildings changing her ways of taking care of us!

As I traveled to my high school reunion last Saturday, I saw a sign that read "Snow Removal" and there was a phone number. I decided that fellow was a real optimist! When the weather is hot, it just pays to use common sense. And remember----- winter is on the way!

BY THE LIGHT OF THE HARVEST MOON

The corn waves on a thousand hills,
Reflected in the sparkling rills;
The earth has had its meed of rain,
The sun has spread its warmth again.
Put in the sickle, reap the corn;
It is the pleasant harvest morn.
(From A Harvest Song by Marianne Farningham)

It was that time of year, we spent days getting the corn cut and shocked --- and a few nights. We would do the evening milking, feed the calves, wash everything up and finish the chores. We would pick up our corn cutters and off to the cornfields we would go, Dad, Mom and I. As we chopped and shocked the corn, we would talk and laugh and discuss the day's happenings. Slowly our light would disappear and up would come that beautiful, full, harvest moon lighting up the fields. We would work until we got tired and then head to the house to get cleaned up and think about bedtime. Some of the days were still very hot and humid but when the sun went down the cornfields would be so cool and working in them was so comfortable. To be trusted to use a very sharp corn cutter and to properly cut and stack the stalks made me feel grownup and proud.

Given the time to dry down, the next big chore would be cornhusking. We all had our own "husking pegs" we liked to use and knew which ones were ours. Dad's was easily recognized, as he had large hands and his peg had a wider leather strap than any of the others. If the weather was promising to be dry for a day or two, everyone headed for the fields and the corn would be husked and left in piles to be picked up later. If we weren't sure what Mother Nature had in store for us, we would get out the wagon we called the "springboard", hitch up the horses and travel to the fields. We would husk out a path into the field, drive the horses and wagon to the center of it, where they would stand and patiently wait until the wagon needed to be moved. The corn would be husked and tossed into baskets to be carried to the wagon. The left over stalks (fodder) would be combined in large shocks to later be hauled off. It was usually thrown out for cattle to eat the blades, if they chose to. Sometimes when the hay crop was short those corn stalks were very helpful in keeping the cattle fed.

There would often be boys or men in the neighborhood seeking extra work and they would be hired to help with the corn cutting or husking. Often those workers gained a "reputation" for how fast they could cut corn or how many bushels of corn they could husk out in a day. There was one gentleman in our neighborhood who was very proud of his speed in corn husking and farmers in the area competed to see who could get him hired first!

It was a good feeling when the corncrib was finally filled. Old corncribs were built with spacing between the boards or slats to allow the air to flow into the crib and help finish drying the corn. When my Dad was feeding it out he didn't want to find any mold! Some of the biggest and best ears were sorted out. They would be put on nails in the granary and saved for seed for planting the next spring. There would be special ears chosen to be thoroughly dried, shelled and ground for cornmeal to make mush, smothered in butter and brown sugar or maple syrup; cornbread; Mrs. Charles Harrison's corn cakes, etc. for the family to enjoy throughout the winter. What a joy it was to see those beautiful golden ears peeping out from inside the corncribs. Another reassurance that there would be survival, for both man and beast, of the cold, harsh, snowy winter on the way. Such a feeling of satisfaction and accomplishment for every farmer!

GET PREPARED, IT'S GOING TO BE A ROUGH WINTER!

Get out your snowplows, make sure you can find your snow shovel, stock up on rock salt, dig out your long johns, Carharts, toboggans, and mittens! It's going to be a rough winter! How do I know? There he was right in front of me as I was mowing the lawn (I hope for the last time)! He must have been at least two inches long, plump, fuzzy, and black. ALL BLACK! The Woolly Bear! He was just wiggling his way along, looking for a place to burrow in for the winter. He was lucky I hit the brake and missed him. He was probably sending out a message to his fellow Woolly Worm friends to "look out for the crazy lady on the green monster, she drives like she is practicing for Daytona"!

According to folklore, the "woolly worm or woolly bear" can predict the kind of winter we will have by his color. If he is brown or mostly brown, it will be a mild winter, but if he is all black, look out, that means a cold snowy, bad winter. The Woolly Bear is actually the caterpillar (larvae stage) of the Isabella Tiger Moth. Found throughout the United States, it is usually seen in the fall looking for the ideal place, usually under rocks or in logs to hibernate. Once spring comes, the larvae will awaken, feed once more, enter the cocoon stage and a few weeks later, the Isabella Tiger Moth will appear. There is a Woolly Worm festival held each year at Lewisburg, Pennsylvania, where worms are captured, measured, and examined. The prediction this year was for a severe winter.

And who hasn't heard, "the higher the hornet's nest, the deeper the snow drifts". I haven't seen a hornet's nest yet this year, but I will be watching for them. Our friendly squirrel,

Rocky, was rushing around the other day gathering nuts and hiding them like crazy. He was even trying to get into the bird house we have on the electric pole outside our kitchen window. He worked for over an hour trying to enlarge the hole. We assume he was looking for a storage unit! His tail was really bushy, waving everywhere in the harsh wind, another sure sign of a hard winter. If he had been in no hurry, it would have indicated a mild winter.

A lot is to be said for conclusions drawn from observing nature. Flowers and plants behave according to weather conditions. The blossoms of both the scarlet pimpernel and the morning glory will open wide, when the sun is about to come out. They will close tightly so as not to damage their delicate petals, when rain is on the way. Tree leaves will turn up before a rain. In coastal areas, seaweed is often used as a natural weather forecaster. Kelp, for example, shrivels and feels dry in fine weather, but swells and becomes damp if rain is in the air. There is an old English saying in the spring, "If the oak flowers before the ash, we shall have a splash. If the ash flowers before the oak, we shall have a soak".

Humans have used animals and witnessed animal behavior to predict weather throughout history, especially before there was science. " Cattle closing together foretells a thunderstorm." Often our cows would come in from pasture early if a storm was brewing, and instead of lying down and chewing their cuds, they would stand in a group at the gate to the barn. My Border Collie, Lass, can tell hours before a storm arrives. She will come to me and stay at my side until the storm arrives and she doesn't leave me until the storm is over.

These are some of the superstitions that have grown up through the years. I don't know if there is any scientific knowledge surrounding these facts, but many people believe them. "When a chicken oils its feathers, you may expect rain". "If a frog croaks more than usual, or toads are seen coming out of their holes in great numbers, or worms are appearing at the surface of the soil, or moles throwing up more soil than usual, or turkeys are collecting together in great numbers, if any of these are seen, then it is a sure sign of heavy rain to come". "A cat turning on its back with its nose up, or cats licking their paws often and rubbing them over their faces is a sure sign of rain." "Restless pigs are said to promise wind as it is said they can actually see the wind". "When ladybugs swarm, expect a day that is warm". "Cobwebs on the grass are a sign of rain." "Ravens croaking loudly in the morning is a sign of a good day". "If the birds gather together in multitudes on the ground there will shortly be snow."

    Before equipment was available people used all the things around them as a guide to the weather, as well as looking at the skies. They used the behavior of animals, plants, and birds as clues to future weather patterns. Come the middle of March, we should know if the Woolly Bear got it right. Believe what you will, but as for me, I am off to town to buy new boots and some wool socks!

OVER THE RIVER AND THROUGH THE WOODS TO GRANDMA'S

We could hardly wait for Christmas Day when I was young, as we knew everyone would be gathering at Grandma and Grandpa's house. Come rain, snow, sleet or ice so bad that one year we had to put chains on the car to get there, we still found a way to somehow get to Grandma's house for Christmas. There would be a big dinner and oh ---the food! Ham, chicken, mashed potatoes with lots of butter on top, Grandma's homemade noodles, corn, Mom's baked apples with caramel sauce and cherry pie, Polly's graham cracker cream pie, cookies of all kinds, Aunt Val's cracker dressing, homemade rolls, Aunt Ruth's chocolate cake, the list goes on! Everyone stuffed themselves until they felt miserable! We all knew that Aunt Georgia would arrive late wearing high heels instead of snow boots and being cold would head straight for the hot air register to stand on!

After the table was cleared and the dishes washed and put away, the adults settled around the kitchen table for card games, euchre and King Pedro, or just relaxed and talked. We kids played various games in the living room and sang songs. Some of the younger ones curled up and took a nap. My family didn't exchange gifts --- none of them had money to buy extra gifts. The kids always received their gifts at home before coming to Grandma's. Usually we all got pajamas, toboggans, mittens, socks, etc. Even though they were things we needed, finding them under the Christmas tree in pretty packages made them special. There would be a toy or two, some oranges and candy. None of our parents could afford a lot of toys and I don't remember that any of us were very disappointed. We understood that we had to be satisfied with the gifts we received. The important thing was all of us getting together at Grandpa and Grandma's, eating, playing with our cousins, having fun together and seeing relatives we hadn't seen for a while. Often those who had cows to milk or animals to take care of would go home to do their chores and then return to party until the early morning hours. Grandma and Grandma didn't have money for gifts, but they gave us lots of hugs and love. Grandpa always managed to have a bag of store bought candy to share with us and that was a special treat.

Those are the things I remember about Christmas when I was young. I don't remember the gifts! I can only remember one or two things I ever received. One was a small blue tractor with a string attached to pull it by and when you did, it went clickity clack, clickity clack. The other was a sled and it never snowed enough after that Christmas for me to sled ride! It wasn't shopping, presents, decorating, and spending money that was important. It was all about the family getting together at Grandma and Grandpa's house on Christmas Day and the good times we had. That is what my cousins and I remember! In today's world there is so much frustration during the holidays. Everyone seems to worry about getting everything just right and buying gifts. Those are not the important things. I hope that families everywhere are getting together to celebrate the real reason for Christmas and to make the memories of family and friends together in peace and harmony. Years later those will be the things remembered. Merry Christmas and Happy New Year to everyone!

ENJOYING BUBBLE AND SQUEAK THIS WEEK

As I write this the sun is shining and it is a beautiful November day. Combines are busy in my area taking off the last soybeans and corn. Hopefully the nice weather will hold until the harvesting is done. Deer hunter s are waiting this week for that big buck to come into shooting range.

Thanksgiving Day was an enjoyable day for many, with family members and friends able to get together to enjoy all the delicious food, the fellowship and love. There was much to be thankful for and the addition of two adorable baby great-grandsons added to the joy in my family. Our hostess on Thanksgiving Day shared the following blessing, " Dear Lord, thank you for all of your blessings, may the stuffing be tasty, may the turkey be plump, may the potatoes and gravy have nary a lump, may the yams be delicious, may the pies take the prize, and may the Thanksgiving dinner stay off of our thighs!"

I was lucky enough to attend a second Thanksgiving dinner in my family on the following Saturday. Both dinners had all the traditional foods as well as many other delicious dishes and desserts. So there was lots of leftovers to share. That can mean it is time for "Bubble and Squeak".

Bubble and Squeak is a traditional English dish made from leftovers from a Sunday dinner or Thanksgiving or Christmas dinner, with the main ingredients being leftover potatoes and leftover vegetables. Often boiled cabbage is the vegetable used, however any cooked vegetable can be added. It was very popular during World War II when food was short due to rationing. Originally a beef-based recipe, history suggests it went to the modern potato based recipe we know today because of the rationing. The earliest recipe was by Maria Eliza Ketelby

Rundell in 1806. It is called Bubble and Squeak because of the noises it makes while frying, though some have said it is the noise your stomach makes after eating it.

There really is no precise recipe because it is always made from the leftovers you have on hand. It is suggested that you try to use equal quantities of potatoes to vegetables, although the potatoes are the most important ingredient, as they bind it together. You can use any type of leftover potatoes. Mash the leftover potatoes in a bowl, if they are too dry an egg can be added. Finely chop the vegetables and mix into the potatoes. Add a thick slice of butter to a non-stick frying pan and when it is hot (don't let it brown) add the mixture from the bowl. Press down with a spatula and smooth it out slightly. If desired you can make patties of the mixture. The mixture can be seasoned with salt and pepper. Leave it undisturbed for five minutes or so or until a nice brown crust forms, then turn it over and repeat before serving. Animal fat can be used or you can fry a little bacon in the pan first, remove it and add to the mixture and then fry the Bubble and Squeak in the bacon fat. Onions can be fried until soft and translucent and added or ham can be added. Bubble and Squeak is often served with a fried or poached egg on top.

There are many ways to use the leftovers from holiday dinners---Bubble and Squeak is just one of them. My grandmother and mother both made it! Use the ingredients you like and enjoy!

GIVE THE SPECIAL GIFT THAT WILL COST YOU NOTHING

As I look back on past holidays, one of my favorite memories is the making of candy with my Dad. After finishing up the milking and barn chores, Dad would fire up the Home Comfort cook stove, and together we would make fudge or taffy. My Dad had lost his mother at two years of age, so in growing up, he and his siblings had to learn to take care of themselves, which included learning how to cook. My Uncle Emmett could bake a three layer chocolate cake that was the envy of ladies everywhere. I could hardly wait for those special evenings and the time spent with my Dad, as well as the joy of the special sweets treats we made together.

The best gift my Dad always gave me, not only at the holidays, but throughout the year was "time". Wherever he went, I was always allowed to "tag along", to the feed store, the machinery dealers, to the neighbors, or working in the fields, and he took time to teach me things. He taught me to love dairy cows, and how to properly care for them, how to drive a team of horses, how to fix fence, how to operate every piece of farm equipment we had. I learned how to work ground and how to work a finishing furrow to his satisfaction, how to mow a hay field so the back swath was properly done, and to rake hay the correct way and how the hay should feel when it was ready to take in. He took time to teach me little things, such as putting tools back where they belonged so when I needed them again I could find them, and to always finish any job I started and do it correctly. He took time to help me with school work, and 4-H projects. When my Dad was older and unable to work, I treasured the time he gave me, sitting on a hay bale and talking to me while I milked the cows.

Time is a precious gift, one that everyone young or old can give, and there are so many ways it can be given. The greatest gift you can give any child is time. Spend time just talking with children, read books to them, play games with them, take time to teach them how to do things. Go to their ball games, musical concerts, 4-H exhibitions, take them for a dish of ice cream or a pizza, or just take a walk with them. They will treasure the time given and I guarantee they will remember it.

Take time to send a lonely person a Christmas card. Visit a nursing home or someone you know who lives alone. Just having someone to talk to gives them so much pleasure. And often they can tell you some of the most interesting stories that you have never heard. Volunteer your time at church, a soup kitchen, or other organizations. Give your time to have a cup of coffee with a friend who is dealing with life's problems and just needs someone to listen. Call up someone and wish them Happy Holidays! Take time to bake someone some cookies. And give yourself some time to read a good book, or watch a movie, or to just relax. There are so many ways to give the gift of time, but in our busy world of today we forget what an important gift it can be. Giving a gift of your time costs nothing, but the rewards can be priceless.

'TWAS THE NIGHT BEFORE CHRISTMAS ON A DAIRY FARM

"Twas the night before Christmas and all through the house not a creature was stirring" ----- It's a little more hectic in some houses. Mom is frantic with getting everything finished up at the last minute. There are still some cookies baking in the oven, presents still need to be wrapped, the scissors are lost, and the dog just ran off with a roll of wrapping paper. Dad is hiding out in the garage trying to make part B fit into slot C after he found out that part U was actually part V.

"The children were nestled all snug in their beds " ----- not quite! They did finally put on their pajamas and some Santa Claus hats and now they are jumping up and down on their beds singing, "Grandma got run over by a reindeer"!

"And Mama in her kerchief and I in my cap"----- Everyone is a lot more comfortable wearing baggy sweat pants, an Ohio State tee shirt, and a hoodie!

"When out on the lawn there arose such a clatter"----- Oh, my gosh, it's Farmer Jones driving his John Deere tractor with blinking Christmas lights all over it and a big star on top of the cab! He's wearing a Santa suit and delivering his wife's "special" fruitcakes to all the neighbors. If only she knew how to cook!

"When what to my wandering eyes should appear" ----- Uh,oh! Grab your coats and boots, that's not eight tiny reindeer out there! The heifers got the gate open again!

"And he whistled and shouted and called them by name" ----- Now Daisy! Now Dolly! And all you other "xyz@&#mq" heifers, get out of the yard and the flower beds!!

"And then in a twinkling I heard on the roof the prancing

and pawing of each little hoof" ----- There they go, right through the neighbor's corn field! They are headed for the road! Get the four wheelers!

"He was dressed in fur from his head to his foot and his clothes were all tarnished with ashes and soot" ----- Sheriff, I am really sorry those heifers knocked you down and got you all muddy. Is your deputy o.k.? Why don't you go into the house and my wife will give you a towel and a cup of coffee and some fruitcake. You might want to check your shoes!

"He spoke not a word but went straight to his work" ----- O.k. boys, take it easy now, lets just drive them slow, keep banging on that bucket of feed, and head for the barn. Easy does it, don't spook them. And when we get them in there, make sure that gate gets latched!

"And laying a finger aside of his nose and giving a nod up the chimney he rose" ----- Lets head up to the house, boys, the wife has lots of hot coffee and fresh baked cookies. I really appreciate all of you coming out to help get those heifers rounded up. Just be careful where you sit down, we still haven't found the scissors!

"And I heard him exclaim as they drove out of sight, "Happy Christmas and to all and to all a good night"! "I have to go check on that new baby in the barn".

FINDING BEAUTIFUL THINGS ON COLD WINTER DAYS

The severe winter weather recently has been very difficult for everyone to deal with. Snow to contend with for several days in a row, below zero temperatures, even more extreme wind chill, winds that drifted the snow and sent the cold through every tiny crack and penetrated every layer of clothing worn. Danger of frostbite, loss of electric power, frozen water lines, tractors that wouldn't start and more. Everyone was fighting the battle of normal winter weather and the depression of those dark, dreary, miserable days and yet if you looked around there was beauty to be found.

At three o'clock in the morning as I wandered around my house, bright eyed and bushytailed, unable to sleep, I checked the outside temperature and it read zero degrees. I walked to my kitchen window, pulled back the curtains and looked out. The three inches of snow that covered the ground looked like it was covered with sparkling diamonds. So beautiful! Even on our worst winter  days, if we just look around, we can find things of beauty. We have days when the trees, bushes, fences, etc. are covered in ice or covered with snow. They look like a work of art completed by Mother Nature and they are breathtakingly beautiful. While we are hoping that it doesn't bring down power lines or branches, we cannot keep from admiring the beauty of it and feeling a little sad when it melts.

Watching as the birds flock to a feeder on the brutally cold days and seeing how much they seem to appreciate finding food available for them is a beautiful sight. You watch them and you can't help but wonder how such tiny creatures can keep from freezing to death in the brutal cold. And where do they find enough shelter to protect themselves from the frigid blowing wind?

You look out and there is a big brown bunny rabbit eating the ear of corn that was put out especially for him. His dark brown fur looks so pretty as he sits in the pure white snow. On a nearby tree a fox squirrel nibbles on his ear of corn surveying the area as he eats. It is said that squirrels hide their collections of nuts for winter in so many different places that they forget where some of them are hid. I don't know if that is true or not, but just in case it is a good idea to help them through the winter by putting out some corn.

It is a beautiful sight to see that big snowplow flying down the road throwing snow everywhere, clearing the way for the milk trucks, emergency vehicles and people who have to get to their places of employment, doctor's appointments, etc. The last few days have been tough ones for our county highway workers and Ohio Department of Transportation employees. They have spent long hours fighting the ice, snow and the drifts that the wind kept blowing back. A beautiful sight for them was when they returned to their garages, cold, tired and hungry, and found crockpots full of hot soup and hot foods sent in by people who appreciated their efforts in trying to keep the roads clear and safe for travel.

Children out building a snowman or fort is always a beautiful sight. They will be laughing and happy, bundled up in layers of warm clothing, mittens, scarfs, and toboggans until all you can see is their noses! They are having fun and enjoying their snow days off from school. Some will be sled riding, others ice skating on the ponds nearby.

Later on in the day I can look out and enjoy the pulchritudinous of my granddaughter's horses as they quietly munch hay from the big bales. They don't seem to be minding the cold as their hair coats are thick and heavy and they have shelter if they want to use it. For me horses are always a beautiful sight.

This time of year the things of beauty are not so obvious. They are there but sometimes you must look to find them and then let them brighten your day. To quote a line from a well-known song by singer and songwriter, Ray Stevens, "Everything is beautiful in its own way". Another well-known quote reads, "It is time to realize that beauty lies in the eye of the beholder".

A BRIGHT RAY OF SUNSHINE ON A COLD JANUARY DAY

The thermometer read a few degrees below zero. The wind was blowing making the wind chill factor much lower. Every school in the area was closed. Cars everywhere refused to start. Courthouses closed, as did businesses in the area. Farmers worked to get tractors started to plow open their lanes and to move hay bales and feed to livestock. Dairy farmers were thawing out water lines, trying to get vacuum pumps running so cows could be milked, and working to get milk trucks in to pick it up. Many didn't make it and huge amounts of milk had to be dumped. When farm wives weren't helping to get chores done, they were making lots of hot soup and keeping the coffee pot on.

There was great concern about the electric supply, with some workers already out working on lines in the bitter, cold weather and others standing by in case they were needed. State transportation workers plowed and salted roads twenty four hours a day. Volunteer workers prepared shelters where people could go to keep warm if necessary. Electric heaters and generators were flying off the shelves at hardware stores. Carharts, sweatshirts, toboggans, ski masks, insulated boots, gloves, and mittens were the fashion of the day.

And then there it was, that bright ray of sunshine on a dark and miserable day, that ray of hope that guaranteed things would get better, delivered by the mailman ("Neither rain nor snow nor dark of night shall keep him from his appointed rounds"),----- the first spring seed and garden catalog! Such a mood changer and spirit lifter on a cold and brutal day!

Look at the pictures of those beautiful flowers –hybrid tea

roses, dahlias, daisies, peonies, and more, the colors so vivid you can almost smell them. Red ripe strawberries, blueberries, raspberries, plums, peaches, sugar sweet cherries, just looking at the pictures makes your mouth water. And the vegetables! Big, red, ripe tomatoes that immediately bring to mind bacon, lettuce, and tomato sandwiches. Rhubarb, oh, the thought of strawberry rhubarb pie -----yum, yum! You can imagine an ear of fresh sweet corn just dripping with butter! They even have orange crisp watermelon. Orange watermelon? Lettuce, cucumbers, radishes, sweet peppers-----a fresh salad would taste so good!

Suddenly spring doesn't seem so far away. After all the days are getting longer! Where are those peat pots left over from last year? We have to have some potting soil. We will need tomato seeds, those Early Girl Hybrids look good, and some Sweet Bell Pepper Mix. It might be fun to plant some of those "blue potatoes" this year. Better get an order in and some plants started. Spring might come early this year. Perhaps surviving those nasty, frigid days is just a case of "mind over matter"!

A GOOD WAY TO START OUT THE NEW YEAR

As I sit here writing on the first Sunday in the New Year, the sun is shining beautifully, the thermometer reads 46 degrees and it is January! There is a saying, "It's not nice to fool Mother Nature", but is Mother Nature trying to fool us? Is this a test to lull us into forgetting what January can really be like and then she will hit us with a normal winter blast? We better stay prepared, there are ten more weeks until spring arrives.

Everyone should have holiday decorations taken down and put away. Many people have already started shopping for next year, purchasing wrapping paper, decorations, Christmas cards and even presents, as the holiday items are all on sale. As we all received Christmas cards from friends and relatives we hadn't heard from for a while, most were cheerful and gave us good news, however there are some with news that deflates the holiday spirit just a bit. It is a part of the holidays that can't be avoided.

I, like most people, own a television set, however the holidays are not an interesting time for me to watch television. For weeks it has been nothing but reruns! Even the ads are reruns! Evidently they think everyone is out Christmas shopping! The movies are all about the holidays – going home for the holidays --- saving something that has to do with the holidays --- falling in love with someone they meet at the holidays --- they are all alike. They count on us forgetting how the story goes! And they rerun the same ones year after year.

This time of year give me a cup of coffee or tea, something to read or a puzzle book. Thank goodness for farm magazines, Holstein and other breed magazines, the local newspapers or a good book. I get especially interested if the Holstein magazine is from years ago! My son found some old Holstein magazines on the internet and purchased them for me. Needless to say, I didn't get much work done the day they arrived! My cousin recently talked about his young great-grandson who loves to read. He would rather read than go out to play and always has a book in his hand. He loves to read about history and can tell about things that have happened years ago in spite of his young age. There is no doubt in my mind that he will grow up to be a very successful young man. There is so much to be learned by reading.

One of the things I received during the holidays was a booklet with exerts from The Farmer's Almanac. It had different types of information for each month of the year. The month of January had information about New Year's traditions concerning food. The tradition in our family over the years has been to always eat sauerkraut and pork on New Year's Day and that would assure us of blessings, good luck and wealth. We just keep on eating pork and sauerkraut and waiting!

There was also a list of natural remedies for getting rid of a cold. Among them was the belief that layers of onions would draw contagious diseases from the patient --- onion poultices. I remember them! This brought to mind neighbors who lived about a mile down the road years ago, Tom and Tiny Pierce. They had retired, moved here from another state, and bought an old house and a few acres. Tom was taking his time remodeling the house and every spring they planted a garden with just about every vegetable and herb you could think of. They also had fruit trees and beautiful flowers. They canned and put up jars and jars of produce from their garden and fruit trees. They followed a very healthful diet every day. Tiny always recommended eating an onion every day, and insisted that if you did you wouldn't catch a cold. We visited often to play canasta with them and I don't ever remember them being sick. They were up in years and still very healthy and active when they decided to leave the area.

So, if I were to recommend a New Year's Resolution, it would be to eat a healthy diet that includes meat, lots of vegetables and an onion every day. Just remember to also include lots of milk and dairy products in that diet! Have a happy, healthy, better New Year!

LET THERE BE LIGHT AND WINDOWS

Sunshine streaming through the window early in the morning ---- a beautiful sight! The recent remodeling project included taking in a section of the back porch and installing a new window. While having a new and more modern bathroom is great, my favorite thing is that new window. Not only does it let in lots of light, it allows me to see things that I missed before. I can see more of my granddaughter's horses as they roam the pasture. There is the quarter horse with the big white blaze in her forehead, and the buckskin, Skip, who tends to be a little mean. The mare, Jesse, is showing her age. They add life and color to an otherwise dull pasture field this time of year. Occasionally I catch sight of the bunny rabbit who lives under the old summer kitchen. I wonder if there will be little ones again this year. A warm spring rain a few days ago brought out fishing worms, one was almost six inches long!

When I sit back and relax in my recliner I can look out the sliding glass doors that lead out on to the deck and see the fields across on the hill. On snowy winter days I can see the tracks left by the deer as they travel down over the hill. They often come down to nibble at the multi-flora rose bushes and the growth in the old apple orchard. Sometimes they will hide there and rest for quite a while. There was a three-legged doe that could always be seen there. There is a huge hawk who circles over, looking for his meal for the day. He has a wide wing spread and at times he is just floating through the air. There are puffy white clouds and beautiful sunsets to be seen.

Even on snowy winter days there are many sights to be seen and enjoyed. The ground can be covered in snow and when the sun comes shining through it will look like it is covered in diamonds. The snow clings to the trees and they look like they have been covered with frosting. The first sign of spring is a little bit of green that shows up in one of the wet spots over on the hill. Before you know it, there will be hay and corn and soybeans growing there. Watching those crops from the time they break through the ground until harvest time can be so enjoyable. You hope for a good season with the needed rain to keep them growing and pray that there won't be any storms to destroy them. The pasture fields along our road are dotted with new baby calves and baby lambs.

With the coming of spring, the birds are returning. The yard was filled with those black birds, (I think they are called grackles) the other day. They were just ambling along plucking something out of the grass, bugs maybe. I saw a pair of blue jays in the maple tree outside the window. The humming birds will be returning before long and I look forward to watching them at my feeders. Sparrows are busy making nests. Leaves will soon be growing on the trees. Crocus and daffodils are blooming everywhere. I still have piles of dirt in my yard that haven't been put back in place yet. On the very top of one pile of dirt is one lone daffodil. It stands blooming proud and tall in all its yellow glory to prove that spring really is here!

There is beauty to be seen every day. Most days it just happens --- on the dark and dreary days we have to look for it. There are so many beautiful and uplifting sights in our country life. How fortunate we are to have them. Take the time each day to treasure them.

HOT COFFEE AND A WARM HOUSE
ON A COLD WINTER DAY

As I sat at my kitchen table the other day trying to come up with a column, the temperature outside was five degrees, the wind was blowing, creating a very low wind chill and looking out my sliding glass door I could see the hills all covered with snow. I was enjoying a hot cup of coffee made in my Mr. Coffee, I had just enjoyed hot soup warmed in my microwave, my house was kept toasty warm by the furnace, and I was dressed in plenty of warm winter clothing. My refrigerator, freezer and cupboard were filled with all kinds of food. As I sipped my coffee, my thoughts turned to my ancestors who, so long ago, settled on this farm. They traveled from Pennsylvania prior to 1840 by ox cart, with their milk cow following behind, to make a new life and a new home for themselves and their family.

There are so many questions that I wish I had the answers for about their journey and their life. I have often wondered, why did they choose to stop here among all these Ohio hills? Was there a special reason or were they just tired of traveling and decided it was time to stop? Did they see things here that made them feel it was a good place to settle? Were there other families settled close by? There was plenty of trees of all kinds for building structures, for wood for cooking and heating a home, and for building items needed inside the home. There was water, a creek close by and several springs. The hills would offer some protection from winter's cold winds and flooding would never be a problem. There was an abundance of wildlife for food, and to make protective clothing and useful items. There were some dangerous animals. The story goes that the man went for supplies and while he was gone the woman had to build fires to keep the wolves from coming in to kill and take the baby calf that had been born.

A log cabin had to be built for them to live in. There was just hatchets, knives, axes, mallets, handsaws, crosscut saws, sledges, chains and hammers to work with, no electric saws  or tools! They had to figure out a way to move and lift logs and stones. They had to fashion a roof and fill the cracks between the logs. There was no mortar or cement, just mud and grass. Did neighbors come in and help them build or did they do it all alone?

They knew winter would be coming. Did they have any idea just how severe winter can be in Ohio? They had to raise some food and find a way to store it for winter. They had to have something to feed their livestock throughout the winter. They couldn't just hop in the four-wheel drive pickup truck and go to town to the grocery store and the feed mill for what they needed. Just getting water for drinking, cooking, and for the livestock had to be difficult. I am sure they didn't take a bath very often!

We have just gone through several days of very cold and very severe winter weather. There have been problems of all kinds to take care of in order to keep our daily lives going. These days have been extremely difficult for dairy farmers and farmers with livestock. Yet we are so fortunate to have all the modern ways to help us handle the problems that this kind of weather throws at us. There had to be winter days when our ancestors feared they might not survive. As I look out at this severe winter weather, I think about what they would have had to do to live through days like this. They had to have tremendous strength, courage and faith. They found the ways to survive the harsh winters and so will we!

LEARN TO ENJOY THE VICISSITUDES IN LIFE

To "tack"--- is to follow a zigzag course. Currently I must tack in my house around boxes, plastic containers, and all sorts of things that are completely out of place. The reason --- remodeling my bathroom. The purpose in doing all this is to "ameliorate" --- make it better or more tolerable. When it is all done, my shower, washer and dryer will be upstairs and my life will be more tolerable because I won't have to go up and down stairs so much.

All of this upheavel has me in a "ferment"--- a state of unrest and disorderly development and it looks like I am going to be that way for a while. I am "discombobulated"---upset and frustrated and it has been a "kerfuffle"---disturbance to my every day routine. I have been told to just enjoy life's "vicissitudes"--- up and downs, as I can't change it anyway. I will just have to "adapt"---make fit, usually by alteration. In other words, get used to it! I will have to put up with the "hiatus"---interruption in time or continuity. All of this has definitely been an interruption in my time, including the time I get up in the morning, since I must be up, dressed, and have my routine in the bathroom done before the workers arrive early. As a retired person that has definitely created a brouhaha---uproar!

All of the pounding, ripping, hammering, sawing, nailing, dust, piles of dirt, mud, and interruptions to my regular schedule does "agitate"---emotionally disturb me. I will do my best to not be "cantankerous"---hard to deal with, as this project goes along. I don't want to wind up in the "hoosegow"---jail for protesting to all of the problems and turmoil. Some days I would just like to "absquatulate"---flee the place. But I am not a "flibbertigibbet"---silly or flighty person, so I will hang in there until this remodeling project is done. I have to endure all these changes in my life just because I am "senescent"---getting old!

There has been a "crescendo"---gradual increase in the problems that have had to be dealt with. Every day we seem to find one more thing that is wrong and needs to be fixed before we can move on. What we thought was a carefully planned out remodeling of the bathroom has turned into a "doozy"---extraordinary one of a kind job! I am "sanguine"---confident the contractor is going to eventually get the job done and that it will be nice. He is not a "blunderbuss"---careless person when it comes to his work. When it is all done I will be filled with "felicity"---happiness. When it is finished and looks so modern and nice, I may just have a "hootenanny"---folk singing event and party to celebrate!

FIND THOSE BRIGHT SPOTS AND ENJOY

A dark, dreary, miserable, cold, windy day, the ground covered with snow, the trees barren of their leaves and there are dark clouds rolling in the sky. The television seems to report only sad, depressing and unsettling news. Surely, somewhere there is happiness and joy, but where do you find them? They are there---we just don't always notice them. Sometimes you really have to look! My son reminded me that the sun is still up there, you just have to get up above those dark clouds as he did on his flight home from Florida. Florida! Temperature 70 degrees! Ohio! Temperature 29 degrees! Bah Humbug!!

As I sit in my easy chair, looking out my sliding glass door, I see brown fields, leafless trees, the summer's bright wild flowers and colors all gone. For the next few months that is all I will see except when the ground is white. There will only be memories of last summer's green alfalfa fields, the tall corn as it tasseled out, the soybeans as they grew thick and green. The three deer I see coming down through those fields are just as brown and blend into the brush in the pasture perfectly.

The hunters will have a hard time seeing them. They may come to visit often. The squirrel I saw climbing around in the tree outside my window this morning was a deep and beautiful reddish color with a big bushy tail. I must get some corn to put out for him, even though there are several walnut trees nearby.

The horses came up from the pasture to hang around this morning. They all have their dark winter coats on. It always gives me joy to look out the window and see them. I always dreamed of owning and riding a beautiful Palomino like Roy Roger's Trigger. My granddaughter's horses are as close as I will ever get to that dream coming true! However, you never know---I have been known to be impulsive and raise my hand at cow sales and buy lottery tickets once in a while! I must stay away from horse sales!

Time out from writing to go get my mail. It is cold outside, so I have to bundle up, but it is so refreshing to breathe in the clean country air! I should go outside more in spite of the cold weather. Mail is something I always look forward to and I seldom have a day that there isn't something in the mailbox. Some days it may just be "junk mail" or bills, but today is a good day as there is my local newspaper and Farmshine. It is a great day when the Ohio Holstein News arrives or Holstein International. I won't get much work done on those days! Reading is one of my favorite things and has been since childhood. My grandfather taught me to read as I sat on his knee at four years of age. I had broken my wrist and couldn't play very well, so he taught me and entertained me at the same time. There is so much to learn, so many places to visit and so many interesting things to enjoy when you read. I like doing crossword puzzles and word search puzzles and I challenge myself by doing them in pen instead of pencil. I recently read that during these dark days you should do them in green or red ink! It feels so good to come back into a warm house, to get a hot cup of coffee, tea, or hot chocolate and sit down and enjoy my mail.

I am lucky to have friends to call or who call me, some are older, some younger, some are retired, some not, some near and some far. Family stops in when they can, but all are busy with work, children, family activities, school, and their daily chores. I know if I need them, they will come. While there is no doubt they too get "down in the dumps" at times, their busy lives do not let them stay there very long! There will be a call this evening from a cousin who is alone. He calls every evening and we discuss the day's happenings, the world's problems, talk about family and reminisce about our childhood and the good old days. We may talk fifteen minutes or we may talk for an hour and a half. The "visit" brightens the day for both of us. This morning there was two unexpected emails from a special friend, one that probably confused a LOT of people besides me and one that made me laugh (maybe I shouldn't have but I did). They made my dark, dreary, cold rainy morning brighter!

Writing my column can be a bright spot for me on a dark dreary day, yet there are times when trying to come up with an idea can be very stressing and it is depressing when I just can't get my brain to work. Then, suddenly, out of nowhere comes a thought, an idea, or I see a word or read an article and the brain starts creating something that I hope the people will read and enjoy.

In each person's life there will be dark and dreary days, but there can also be bright spots in those days that we don't recognize or that we just take for granted as part of our day. Everyone has to deal with those days along with the daily problems. My dark and dreary day is what I choose to make it. I can sit around feeling sad, lonely, and sorry for myself or I can appreciate the little things that help to make the day a little brighter and I can choose to do things that make the day more enjoyable. The choice is mine. Today my choice was writing a column. I hope it will brighten someone's day and bring joy.

www.ingramcontent.com/pod-product-compliance
Lightning Source LLC
Chambersburg PA
CBHW070850050426
42453CB00012B/2118